Plasmonic Resonators

Plasmonic Resonators

Fundamentals, Advances, and Applications

Masanobu Iwanaga

Pan Stanford Publishing

Published by

Pan Stanford Publishing Pte. Ltd.
Penthouse Level, Suntec Tower 3
8 Temasek Boulevard
Singapore 038988

Email: editorial@panstanford.com
Web: www.panstanford.com

British Library Cataloguing-in-Publication Data
A catalogue record for this book is available from the British Library.

Plasmonic Resonators: Fundamentals, Advances, and Applications

ISBN 978-981-4745-06-2 (Hardcover)
ISBN 978-981-4745-07-9 (eBook)

Printed in the USA

To my family

with thanks for longtime support

Contents

Preface

Plasmonic resonators, composed of metallic micro- and nanostructures, belong to category of excited-state physics on resonances at gigahertz to petahertz. Dynamical physics is in contrast to ground-state physics that includes, in a wide sense, thermal states.

Considering the above feature, this book has three main aims.

1. To provide fundamental knowledge on plasmonic resonators. Optical properties in metals, Maxwell equations, and fundamental physical theory on resonances, i.e., response function theory are described. In addition, the history of plasmonic resonators is addressed.

   Chapters 1, 2, and 4 are mostly devoted to this purpose.

   This part will be useful for students.
2. To convey information about the diverse plasmonic resonators. The field of plasmonic resonators is already quite developed, and this fact seems to suggest that there is not much room left to find new types of plasmonic resonators. Therefore, this book is most likely enough to know the possible types of plasmonic resonators.

   Chapter 3 is responsible for this purpose and is based on experimentally examined plasmonic resonators.

   This part will be informative for a wide range of readers.
3. To stimulate further development on plasmonic resonators. Plasmonic resonators have already attained several examples realizing significant plasmon-enhanced effects.

   Chapter 5 mainly addresses experimental results, which are more than the simple electric-field enhancement and are based on more in-depth strategies. This direction, I expect, will enable researchers to achieve substantial progress in the next 5 years or so.

Chapter 6 provides a summary and discusses the future prospects related to plasmonic resonators.

This part will be a starting point to (near) future establishments on plasmonic resonators.

More specifically, this book is organized as follows.

- Chapter 1 describes the basics of plasmonic resonators such as optical properties of metals and surface plasmon polariton and surveys the history of plasmon studies in which metallic thin films, gratings, extraordinary transmission, and metamaterials appear.
- Chapter 2 is devoted to response functions in view of both classical and quantum mechanics. Configuration-interaction theory is also addressed with some analyses based on the Fano resonances.
- Chapter 3 addresses the various plasmonic resonators that have been found to date. A new class of plasmonic resonators, termed stacked complementary (SC) plasmonic resonators, is also included.
- Chapter 4 is devoted to nonlocal responses by metal and plasmons, and to optical nonlocality in plasmonic resonators.
- Chapter 5 describes the recent advances in various plasmonic enhancement, designated as Plas*, and in a few applications.
- Chapter 6 is devoted to future perspectives, which are awaiting challenges in the near future.

I hope that readers will grasp the basics of plasmonic resonators and that this book will help the readers contribute to the diverse applications in the near future. The book does not provide the *conclusions* on plasmonic resonators but intends to stimulate further advances based on the progress to date. On finishing this manuscript, I realize that there is much room for plasmonic resonators to develop.

Masanobu Iwanaga
Tsukuba, Japan
June 2016

Acknowledgments

The author is grateful to the colleagues in NIMS[a] for daily discussions. In particular, I would like to thank Dr. Bongseok Choi for his contributions to the nanofabrications of plasmonic resonators; some of the experimental results in this book are owing to his contributions.

I also acknowledge financial supports by grants: JSPS[b] KAKENHI (Grant numbers 26706020, 26600121), MEXT[c] KAKENHI (Grant number 22109007), JST[d] PRESTO Program, NIMS 3rd mid-term research project "Innovative Photonic Materials," and Seeds Development Research Grants in NIMS. In addition, I am grateful for the support in numerical implementations by Cyberscience Center, Tohoku University and Cybermedia Center, Osaka University through HPCI[e] System Research Projects (IDs: hp120066, hp140068, hp150043) in Japan.

---

[a] National Institute for Materials Science, Japan.
[b] Japan Society for the Promotion of Science.
[c] Ministry of Education, Culture, Sports, Science and Technology, Japan.
[d] Japan Science and Technology Agency.
[e] High Performance Computing Infrastructure.

Chapter 1

Introduction

Contemporary optical science consists of several subfields as summarized in Fig. 1.1. The subfields can be categorized by wavelengths of electromagnetic (EM) waves and the scale that is the periodicity or representative dimension of objects.

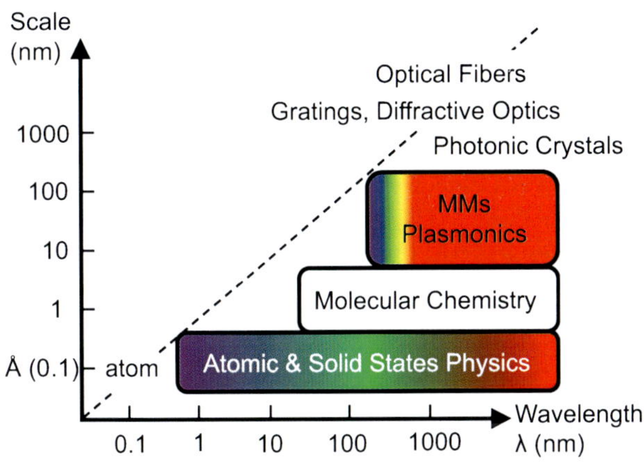

Figure 1.1 Optical science diagram: scale versus wavelength. The scale denotes the periodicity or representative dimension of the objects. The dashed line represents the boundary where scale = wavelength.

Plasmonic Resonators: Fundamentals, Advances, and Applications
Masanobu Iwanaga

ISBN 978-981-4745-06-2 (Hardcover), 978-981-4745-07-9 (eBook)
www.panstanford.com

Broadly speaking, atomic and solid states physics and molecular chemistry are natural sciences, located at the small scale that is less than 10 nm. The wavelengths used for atoms and solids spread over a very wide range from the x-ray to far infrared (IR).

Other subjects such as optical fibers, gratings, diffractive optics, photonic crystals, and optical metamaterials (MMs) and plasmonics are mostly located at the scale larger than that of the natural sciences. Thus, these artificially structured objects fill a large area, being complementary to the natural sciences. We are mainly going through the domain occupied by MMs and plasmonics in this book from the viewpoint of plasmonic resonators that are the key elements.

In this chapter, we first survey the physical quantities that characterize metals such as plasma frequency in Section 1.1 and optical constants in metals in Section 1.2. Related to the optical constant, Drude-type permittivity metals are described and the actual permittivity in metals is summarized based on measured data. The metal–insulator interface is described, where surface plasmon polaritons (SPPs) emerge in Section 1.3. The history of plasmonic resonators is briefly outlined in Section 1.4. The basic tools to compute plasmonic resonators are reviewed in Section 1.5. Nanofabrication to realize the plasmonic resonators is looked over in Section 1.6.

1.1 Plasma Frequency: Concept of Bulk Plasmon

Plasmons were originally introduced as an elementary excitation of plasma, which is collective excitation of free electrons or ionized atoms in the classical framework as illustrated in Fig. 1.2.

Let us derive the plasma frequency shortly. When one assumes ionized particles of the number density N in a unit volume, an externally applied electric (E) field $\mathbf{E}_{\text{ext}}$ produces the electric displacement $\mathbf{D}$ such that

$$\mathbf{D} = \mathbf{E}_{\text{ext}} + \mathbf{P}_{\text{ind}} \tag{1.1}$$

where $\mathbf{P}_{\text{ind}}$ denotes polarization induced by driven charges. When $\mathbf{E}_{\text{ext}} = E_x\hat{\mathbf{x}}$ ($\mathbf{x}$: unit vector parallel to the x axis), the induced $\mathbf{P}_{\text{ind}}$ is

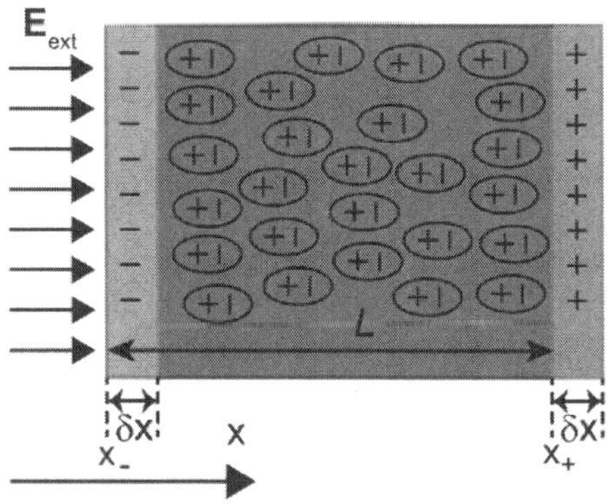

Figure 1.2 Schematic of bulk plasma. External electric field $\mathbf{E}_{\text{ext}}$ induces unbalanced charges $+$ and $-$ at domain length L.

dipole moment per unit volume and given by

$$\mathbf{P}_{\text{ind}} = -\frac{eN(\delta x)SL}{SL}$$

where e denotes the charge of electron ($e > 0$), δx displacement induced by the $\mathbf{E}_{\text{ext}}$, S dimension of section perpendicular to the x axis, L length of domain. Once the $\mathbf{E}_{\text{ext}}$ is switched off, electric field in the domain is given by

$$\mathbf{E} = \frac{\mathbf{P}_{\text{ind}}}{\varepsilon_0} = -\frac{eN(\delta x)}{\varepsilon_0}$$

where ε_0 is the permittivity in vacuum. Then, Newtonian equation of motion for the carriers is given by

$$e\mathbf{E} = m\frac{d^2\mathbf{x}}{dt^2} \tag{1.2}$$

where m is mass of charged particles. The positive (or negative) charges are located at $x = x_+ + \delta x$ (or $x = x_- + \delta x$) and then the solution is a harmonic oscillation such as $\delta x \propto \exp(i\omega t)$. The equation is reduced such as

$$\frac{Ne^2}{\varepsilon_0} = m\omega^2.$$

Thus, resonant frequency, which is named plasma frequency ω_p, is obtained such that

$$\omega_p = \sqrt{\frac{Ne^2}{\varepsilon_0 m}}. \tag{1.3}$$

The plasma oscillation at ω_p is a collective excitation, which is usually called bulk plasmon. This model for the bulk plasmon was first introduced in 1929 (Tonks and Langmuir, 1929) and has been a standard model to described the plasma frequency in the textbooks (for example, Wooten, 1972; Ashcroft and Mermin, 1976; Kittel, 2004). Note that conductive electrons in metals are often approximated as free electrons; therefore, the bulk plasmons at ω_p are also considered in metals.

Quantitative estimation Let us estimate the plasma frequency for ionized plasma and plasma in metal. A typical density in ionized plasma prepared easily in laboratory has $N = 10^{15}$ cm^{-3} and the ions have about $m = 1000m_e$ where m_e denotes the free-electron mass. On the other hand, in metal, the free-electron density N_e is typically $N_e = 10^{22}$ cm^{-3} and we assume that the effective mass is $10m_e$; from Eq. 1.3, we obtain an estimated ratio,

$$\omega_p/\omega_p^{(\mathrm{M})} \sim 10^{-6}, \tag{1.4}$$

where ω_p is plasma frequency of the ions and $\omega_p^{(\mathrm{M})}$ is plasma frequency in metal.

In the estimation, the plasma frequency in metal $\omega_p^{(\mathrm{M})}$ was found to be tens of electron volt (eV), which is in fairly good agreement with the actual $\omega_p^{(\mathrm{M})}$ in Ag, Au, Al, etc. in Table 1.1.

From Eq. 1.4, the plasma frequency ω_p in ionized plasma ω_p resulted in GHz range; the value is consistent with conventional plasma experiment conducted with GHz EM waves.

Table 1.1 List of plasma frequency ω_p (Rakić et al., 1998)

Metals	Plasma frequency ω_p (eV)	ω_p ($\times 10^{15}$ Hz)
Ag	9.01	13.69
Au	9.03	13.72
Al	14.98	22.76
Cu	10.83	16.45
Pt	9.59	14.57
Ti	7.29	11.08
Ni	15.92	24.19

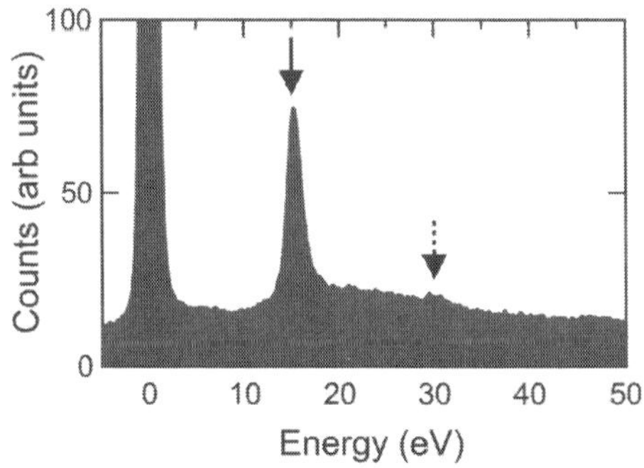

Figure 1.3 Measured EELS of Al. Solid-line arrow at 15 eV indicates the energy of the plasma frequency ω_p, whose secondary peak is indicated by dotted arrow. Owing to the courtesy by Dr. K. Mitsuishi (NIMS).

Experimental observation of the bulk plasmon There are mainly two experimental methods to determine the plasma frequency ω_p: One is a direct method to measure electron energy-loss spectrum (EELS), which uses electron beam; the other is ellipsometry that measures permittivity of metals in optical manner and to determine the ω_p, which is described in Section 1.2. The EELS is mainly used at large energy range more than 10 eV though it is, in principle, possible to use it at a few eV. The ellipsometry is a standard method in the energy range less than 5 eV.

Figure 1.3 shows a typical EELS of Al, measured up to 50 eV. The Al sample was a slice cut out from Al film; the film was made by sputtering on the quartz substrate. The thickness of the slice was about 100 nm. A prominent peak at 15 eV indicated by solid-line arrow comes from bulk plasmon in Al. The plasma frequency ω_p of Al is in good agreement with the ω_p that was obtained based on optical measurement in Table 1.1. Besides, small peak at 30 eV indicated by dotted arrow corresponds to the secondary excitation of the bulk plasmon. The signal at 0 eV corresponds to the zeroth-order electron beam (or the energy beam without energy loss).

We mention that multiple excitations of the bulk plasmons up to the 5th order were observed in EELS measurement for Al foil (Marton et al., 1962), which was approximately 4 times thinner than the specimen in Fig. 1.3.

1.2 Optical Constants in Metals

Metals are characterized for light with optical constants. Generally, materials parameters for EM waves are complex refractive index $\hat{n}(= n + i\kappa)$, relative permittivity ε, and relative permittivity μ. The ε and μ appear in Maxwell equations (Eqs. 1.7 and 1.8), and are related to $\hat{n}$ as

$$\hat{n} = \sqrt{\varepsilon\mu}. \tag{1.5}$$

Note that the parameters ε and μ are generally complex-valued. Here we mainly think of phenomena at optical frequencies and of non-magnetic metals; accordingly, relative permittivity μ is assumed to be unity. Equation 1.5 is simplified as

$$\hat{n} = \sqrt{\varepsilon}, \tag{1.6}$$

which also holds for non-magnetic materials in general. We from now on use Eq. 1.6 as a basic relation.

1.2.1 *Electromagnetic Responses and Constants*

Maxwell equations at high frequencies are expressed in a standard form (for example, Jackson, 1999) by assuming that genuine charge and current do not exist.

$$\nabla \times \mathbf{E}(\mathbf{r}, t) = -\frac{\partial \mathbf{B}(\mathbf{r}, t)}{\partial t} \tag{1.7}$$

$$\nabla \times \mathbf{H}(\mathbf{r}, t) = \frac{\partial \mathbf{D}(\mathbf{r}, t)}{\partial t} \tag{1.8}$$

$$\nabla \cdot \mathbf{D}(\mathbf{r}, t) = 0 \tag{1.9}$$

$$\nabla \cdot \mathbf{B}(\mathbf{r}, t) = 0 \tag{1.10}$$

$\mathbf{E}$ is E-field vector, $\mathbf{H}$ is magnetic-field vector, $\mathbf{D}$ is electric flux density, and $\mathbf{B}$ is magnetic flux density. Variables $\mathbf{r}$ and t denote spatial and temporal coordinates, respectively.

In addition, constitutive equations for local and non-chiral media are written such that

$$\mathbf{D}(\mathbf{r}, t) = \varepsilon_0\varepsilon\mathbf{E}(\mathbf{r}, t), \tag{1.11}$$

$$\mathbf{B}(\mathbf{r}, t) = \mu_0\mu\mathbf{H}(\mathbf{r}, t), \tag{1.12}$$

where μ_0 denotes permittivity in vacuum. The ε and μ generally include response of material to the applied EM fields, and are directly related to response functions as described in Chapter 2. Note that ε and μ are tensors for isotropic, uniaxial, or biaxial media.

When we consider a general case to use a monochromatic plane wave, EM vectors $\mathbf{E}(\mathbf{r}, t)$ and $\mathbf{H}(\mathbf{r}, t)$ are expressed as

$$\mathbf{E}(\mathbf{r}, t) = \mathbf{E}\exp(i\mathbf{k}\cdot\mathbf{r} - i\omega t), \tag{1.13}$$

$$\mathbf{H}(\mathbf{r}, t) = \mathbf{H}\exp(i\mathbf{k}\cdot\mathbf{r} - i\omega t), \tag{1.14}$$

where $\mathbf{k}$ denotes wavevector and ω is angular frequency. Then, Eqs. 1.7 and 1.8 are modified by using Eqs. 1.11 and 1.12 in the following:

$$\mathbf{k}\times\mathbf{E} = \omega\mu_0\mu\mathbf{H}, \tag{1.15}$$

$$\mathbf{k}\times\mathbf{H} = -\omega\varepsilon_0\varepsilon\mathbf{E}. \tag{1.16}$$

Figure 1.4 shows a simple optical configuration where an incident plane wave illuminates the boundary at the normal. As an explicit case, the EM polarization vectors, defined in Eqs. 1.13 and 1.14, for x-polarized EM wave traveling along the z axis (i.e., $\mathbf{k}\|z$) are written as

$$\mathbf{E} = \begin{pmatrix} E_x \\ 0 \\ 0 \end{pmatrix}, \quad \mathbf{H} = \begin{pmatrix} 0 \\ H_y \\ 0 \end{pmatrix}. \tag{1.17}$$

From Eq. 1.16, the H_y component is represented with E_x:

$$k_z H_y = -\omega\varepsilon_0\varepsilon E_x. \tag{1.18}$$

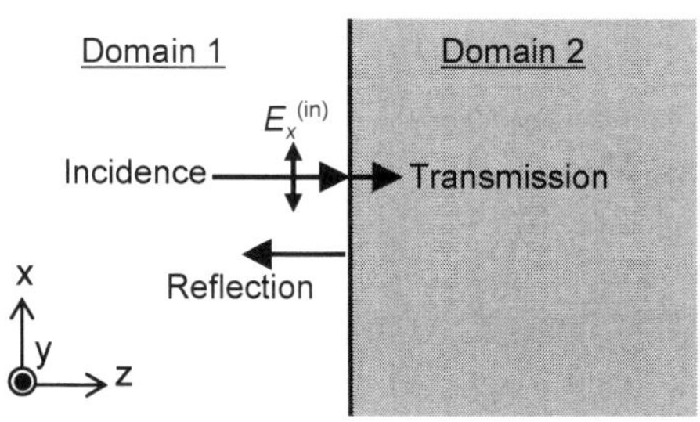

Figure 1.4 A simple optical configuration where only one boundary (black vertical line) exists at the interface of the two semi-infinite domains.

The boundary conditions, that is, $E_x^{(\mathrm{in})} + E_x^{(\mathrm{r})} = E_x^{(\mathrm{t})}$ and $H_y^{(\mathrm{in})} + H_y^{(\mathrm{r})} = H_y^{(\mathrm{t})}$, are expressed in a matrix form:

$$\begin{pmatrix} 1 & 1 \\ k_z^{(1)}/\varepsilon_1 & -k_z^{(1)}/\varepsilon_1 \end{pmatrix} \begin{pmatrix} E_x^{(\mathrm{in})} \\ E_x^{(\mathrm{r})} \end{pmatrix} = \begin{pmatrix} 1 & 1 \\ k_z^{(2)}/\varepsilon_2 & -k_z^{(2)}/\varepsilon_2 \end{pmatrix} \begin{pmatrix} E_x^{(\mathrm{t})} \\ 0 \end{pmatrix}. \tag{1.19}$$

Therefore, Eq. 1.19 is modified such that

$$\begin{pmatrix} E_x^{(\mathrm{in})} \\ E_x^{(\mathrm{r})} \end{pmatrix} = \begin{pmatrix} 1 & 1 \\ k_z^{(1)}/\varepsilon_1 & -k_z^{(1)}/\varepsilon_1 \end{pmatrix}^{-1} \begin{pmatrix} 1 & 1 \\ k_z^{(2)}/\varepsilon_2 & -k_z^{(2)}/\varepsilon_2 \end{pmatrix} \begin{pmatrix} E_x^{(\mathrm{t})} \\ 0 \end{pmatrix}. \tag{1.20}$$

Reflectance (R) is defined by

$$\mathrm{R} = \left| \frac{E_x^{(\mathrm{r})}}{E_x^{(\mathrm{in})}} \right|^2. \tag{1.21}$$

If one uses a relation that $k_z^{(i)} = \hat{n}_i k_0$ ($\hat{n}_i$: complex refractive index of domain i, k_0: wavenumber in vacuum, and $i = 1, 2$) and assumes Eqs. 1.5 and 1.21 results in

$$\mathrm{R} = \left| \frac{\sqrt{\mu_1/\varepsilon_1} - \sqrt{\mu_2/\varepsilon_2}}{\sqrt{\mu_1/\varepsilon_1} + \sqrt{\mu_2/\varepsilon_2}} \right|^2. \tag{1.22}$$

Note that $\sqrt{\mu_i/\varepsilon_i}$ is impedance in the domain i. Obviously, when the impedances of the domains 1 and 2 are equal, $\mathrm{R} = 0$; the incidence perfectly transmits at the interface of the domains 1 and 2. This situation is called impedance matching.

If one further assumes Eq. 1.6, Eq. 1.22 is simplified as

$$\mathrm{R} = \left| \frac{\hat{n}_1 - \hat{n}_2}{\hat{n}_1 + \hat{n}_2} \right|^2. \tag{1.23}$$

As is shown in this example, optical responses are directly in connection to optical constants. In reverse, it is possible to extract optical constants from optical responses such as R; the generalized method for the extraction has been well established as ellipsometry or polarimetry. Practically, incident angles and wavelengths are varied in a sequence of ellipsometry procedure to determine the values of complex refractive index. In single crystals, which have the principal axes and anisotropy in general, the ellipsometry procedure becomes complicated. In contrast, metals, which are mainly addressed in this book, are made by thin film making techniques and random-oriented polycrystalline, so that they can be treated as isotropic media. As for the single metallic crystals, we later describe the nonlocal responses in Chapter 4.

1.2.2 *Drude Metal*

Drude metal describes qualitative properties of free-conducting electrons. In this subsection, the basics are addressed.

Permittivity of Drude metal ε is given by

$$\varepsilon(\omega) = \varepsilon_\infty - \frac{\omega_p^2}{\omega(\omega + i\gamma)} \tag{1.24}$$

where ω is angular frequency, ε_∞ permittivity at higher frequency limit ($\omega \to \infty$), and γ damping constant representing optical loss. Usually the ε_∞ is set to be $\varepsilon_\infty = 1$ because electrons and atoms cannot respond to the applied EM field at the higher frequency limit; we thus set $\varepsilon_\infty = 1$ from now on.

Derivation of Drude permittivity Let us briefly derive the expression of Eq. 1.24. First, we assume that electrons freely stay in a monochromatic time-dependent E field such as

$$\mathbf{E}(t) = \mathrm{Re}[\mathbf{E}(\omega)\exp(-i\omega t)]. \tag{1.25}$$

In the classical framework, an electron follows the following equation of motion:

$$\frac{d\mathbf{p}(t)}{dt} = -\frac{\mathbf{p}(t)}{\tau} - e\mathbf{E}(t) \tag{1.26}$$

where $\mathbf{p}(t)$ denotes the electron momentum, τ relaxation time of the electron, and the charge $e > 0$. Here we think of a harmonic solution and assume that

$$\mathbf{p}(t) = \mathrm{Re}[\mathbf{p}(\omega)\exp(-i\omega t)]. \tag{1.27}$$

Substituting Eqs. 1.25 and 1.27 into Eq. 1.26, one has the following relation of $\mathbf{p}$ and $\mathbf{E}$:

$$\mathbf{p}(\omega) = -\frac{e\mathbf{E}(\omega)}{1/\tau - i\omega} \tag{1.28}$$

When we write the number density of the electrons as N and the mass of electron as m, current density $\mathbf{j}(\omega)$ at ω is expressed as

$$\mathbf{j}(\omega) = -\frac{Ne\mathbf{p}(\omega)}{m}. \tag{1.29}$$

The $\mathbf{j}$ has another expression such as

$$\mathbf{j}(\omega) = \sigma(\omega)\mathbf{E}(\omega) \tag{1.30}$$

using conductivity $\sigma(\omega)$. Equation 1.30 is usually called Ohm law. From Eqs. 1.29 and 1.30, the frequency-dependent conductivity $\sigma(\omega)$, which is often called optical conductivity, is written as

$$\sigma(\omega) = \frac{\sigma_0}{1 - i\omega\tau} \tag{1.31}$$

where $\sigma_0 = Ne^2\tau/m$, which becomes equal to the conductivity σ_0 at the low frequency limit ($\omega \to 0$).

When genuine current exists, one of Maxwell equations is expressed as

$$\nabla \times \mathbf{H}(t) = \mathbf{j}(t) + \frac{\partial\{\varepsilon_0\mathbf{E}(t)\}}{\partial t}. \tag{1.32}$$

Note that Eq. 1.8 represents the case of $\mathbf{j} = \mathbf{0}$. Recalling the time dependence of $\exp(-i\omega t)$, a wave equation represented with E-field vector $\mathbf{E}$ is expressed as

$$-\nabla^2\mathbf{E}(\omega) = \left(\frac{\omega}{c_0}\right)^2 \left(1 + \frac{\sigma(\omega)}{\omega}\right) \mathbf{E}(\omega) \tag{1.33}$$

where c_0 denotes the velocity of light in vacuum. On the other hand, $-\nabla^2\mathbf{E}$ is written as

$$-\nabla^2\mathbf{E}(\omega) = \left(\frac{\omega}{c_0}\right)^2 \varepsilon(\omega)\mathbf{E}(\omega) \tag{1.34}$$

using Eqs. 1.6 and 1.13, and $k_0 = \omega/c_0$. Thus, one reaches the relation of

$$\varepsilon(\omega) = 1 - \frac{\omega_p^2}{\omega(\omega + i/\tau)} \tag{1.35}$$

where ω_p is the plasma frequency in Eq. 1.3. We here define damping constant γ by $\gamma = 1/\tau$; substituting the γ into Eq. 1.35, we reach an equation same to Eq. 1.24 under the assumption of $\varepsilon_\infty = 1$.

Qualitative behavior of Drude permittivity Equation 1.24 or 1.35 is rewritten with separating the real and imaginary parts:

$$\varepsilon(\omega) = 1 - \frac{\omega_p^2}{\omega^2 + \gamma^2} + i\frac{\omega_p^2 \gamma}{\omega(\omega^2 + \gamma^2)}. \tag{1.36}$$

When we think of the low-frequency limit, that is, the ratio $|\omega/\gamma| \ll 1$, Eq. 1.36 is approximated as

$$\varepsilon(\omega) \approx 1 - \frac{\omega_p^2}{\gamma^2} + i\frac{\omega_p^2}{\gamma\omega}. \tag{1.37}$$

Equation 1.37 implies that

$$\mathrm{Re}(\varepsilon) \approx \text{constant} \tag{1.38}$$

$$\mathrm{Im}(\varepsilon) \propto \omega^{-1} \tag{1.39}$$

at the low-frequency limit. This property is confirmed later in actual metals (Figs. 1.5 and 1.6).

Strictly, ideal Drude metal does not exist. This is because free electrons in actual metals are, strictly, not free for EM waves in all the frequency range. Instead, interband transitions of the electrons in metals strongly modify behaviors of electrons at the frequency range far below the plasma frequency and deviate the permittivity from the Drude type. This deviation becomes prominent particularly at optical frequencies ($\gtrsim 10^{15}$ Hz).

1.2.3 *Actual Metals: Comparison of Ag, Au, Al, . . .*

Plasmonic resonators are usually produced using Drude-like metals. We here describe typical Drude-like metals such as silver (Ag), gold (Au), and aluminum (Al), including the interband transitions.

Lorentz-Drude (LD) model In experiment, plasmonic resonators usually include thin metallic elements, which are made by thin-film making techniques such as vacuum evaporation and sputtering. As a result, the thin metal elements become polycrystals or assembles of nanocrytals. The permittivity[a] shown below was also measured using the thin-film metals, representing the polycrystalline properties. The permittivity is assumed to represent local response and

[a]Although it is strictly relative permittivity, we simply call it permittivity from now on, because confusion does not take place.

to be independent of wavevector $\mathbf{k}$. Note that, when one studies single crystals of the metals, spatial dispersion in permittivity and crystalline axes should carefully be taken into account. In this book, results on the single crystals appear in Section 4.1.

The permittivity of metals was often approximated by the sum of Drude- and Lorentian-type permittivity as follows:

$$\varepsilon(\omega) = 1 - \frac{f_p\omega_p^2}{\omega(\omega + i\gamma)} + \sum_{k=1}^{n} \frac{f_k\omega_p^2}{\omega^2 - \omega_k^2 - i\gamma_k\omega} \tag{1.40}$$

where f_k is oscillator strength and the suffix k runs over all the contributing oscillators (or interband transitions of electrons in metals). As we mentioned above, we here assume that the permittivity is independent of $\mathbf{k}$. Note that f_p is added as oscillator strength at ω_p, represents the finite probability, and is not taken into account in Eq. 1.24. Equation 1.40 provides fairly good approximation to the measured permittivity. In particular, it is enough good at the range of $\omega \leq 10^{13}$ Hz.

Brendel-Bormann (BB) model A better approximation was given by use of BB model at the optical frequency, as shown in the following formula (Rakić et al., 1998):

$$\varepsilon(\omega) = 1 - \frac{f_p\omega_p^2}{\omega(\omega + i\gamma)} + \sum_{k=1}^{n} \chi_k(\omega) \tag{1.41}$$

$$\chi_k(\omega) = \int_{\infty}^{+\infty} \frac{dx}{\sqrt{2\pi}\sigma_k} \exp\left(-\frac{(x-\omega_k)^2}{2\sigma_k^2}\right) \frac{f_k\omega_p^2}{(x^2-\omega^2) - i\gamma_k\omega} \tag{1.42}$$

where n denotes the number of the electronic transitions to be taken into account. Equation 1.41 means that the Lorentz oscillators contributing to a transition at $\omega = \omega_k$ are superimposed and have effective line width σ_k due in part to inhomogeneity in actual metals. Equations 1.41 and 1.42 provide a better approximation to the experimental data around the optical frequencies; the calculated permittivity is shown in Fig. 1.5 where the ω_p was taken from Table 1.1. The function $\chi_k(\omega)$ in Eq. 1.42 is expressed as a sum of complementary error function [erfc(z)] or confluent hypergeometric function named Kummer function [$U(1/2, 1/2, z^2)$] where z denotes complex number.

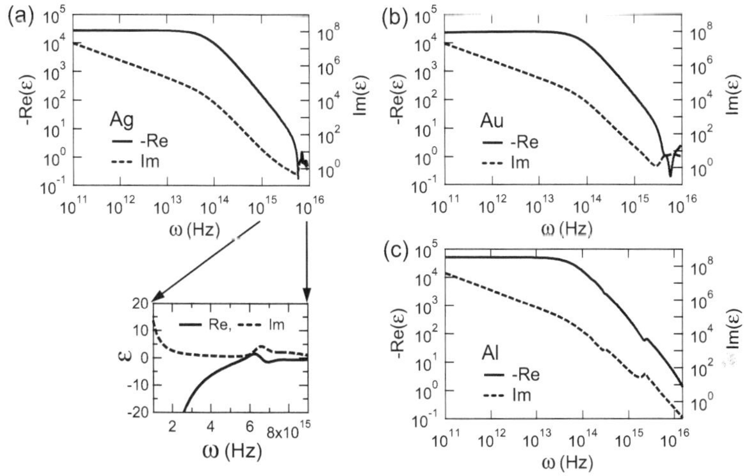

Figure 1.5 Permittivity approximated by the BB model in Eq. 1.41, represented in the log-log scale. (a) Ag. The small panel below shows an enlarged view of the ε in the UV range with the linear scale. (b) Au. (c) Al. The real parts are shown with solid lines and the imaginary parts with dashed lines. Note that the negative-valued real parts are multiplied by -1 to fit the log-log representation.

Figure 1.5 shows permittivity of Ag, Au, and Al. Note that the real parts are shown with multiplying Re(ε) by -1 to fit the log-log scale. A feature of Drude metal is seen at the frequency range of $\omega \leq 10^{15}$ Hz, where the Re(ε) takes large negative values, close to constant as the frequency gets smaller ($\omega \lesssim 10^{13}$ Hz) as shown in Eq. 1.38, and the Im(ε) continuously gets larger in the wide range as shown in Eq. 1.39. Thus, permittivity of metals is qualitatively well described by Drude model (Eq. 1.36).

The permittivity of Ag and Au is not purely Drude type because Ag and Au single crystals have interband transitions between the *d*–*d* states at the UV and visible ranges, respectively (Wooten, 1972). As for Ag, the direct transition corresponds to the peak of Im(ε) at 6.5×10^{15} Hz (i.e., 4.3 eV). Au has the direct transition around 3.5 $\times$ 10^{15} Hz (i.e., 2.3 eV). Due to the direct transitions, Ag and Au have excited electrons via the interband transitions at the frequency and the higher frequencies; as a result, excited (or hot) electrons are generated and the free-electron model is not applicable to Ag

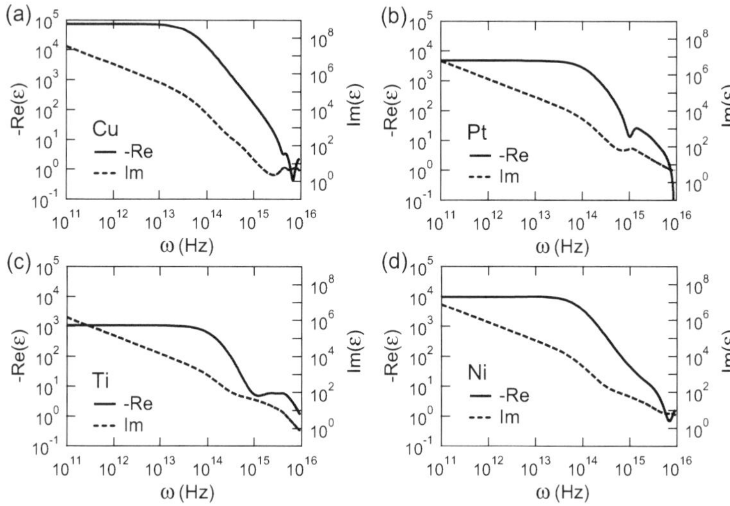

Figure 1.6 Permittivity approximated by the LD model in Eq. 1.40, represented in the log-log scale. (a) Cu. (b) Pt. (c) Ti. (d) Ni.

and Au at the frequency ranges. In contrast, Al does not have strong interband transition at the visible and UV ranges; therefore, Al keeps Drude-metal-like property even at the UV range, having relatively lower loss than Ag and Au at 10^{16} Hz.

The measured permittivity of Ag etc. was obtained by ellipsometry. By measuring polarized reflection of the thin-film metals with varying the incident angles, the permittivity is evaluated. In principle, analysis for Eq. 1.23 enable to extract $\hat{n}$ from the optical responses; if one further assumes that $\mu = 1$, which is valid at optical frequencies (Landau et al., 1984), ε is evaluated from Eq. 1.6.

Figure 1.6 shows permittivity of copper (Cu), platinum (Pt), titanium (Ti), and nickel (Ni), which was evaluated by the LD model. As we mentioned, the LD model is less precise at the visible range than the BB model but provides a good approximation over the wide range. Pt, Ti, and Ni also have Drude-like permittivity at $\omega < 10^{14}$ Hz whereas they have large imaginary part at $\omega = 10^{15}$ Hz than Ag, Au, and Al in Fig. 1.5. Therefore, Pt, Ti, and Ni are known as lossy metal at the wavelengths shorter than the near infrared (NIR).

At the end of this subsection, let us refer to individual features of Cu, Pt, Ti, and Ni.

- Cu has permittivity similar to Ag and Au at $\omega < 10^{14}$ Hz, reflecting the fact that Cu is good conductor. Therefore, in addition to the extensive usage as electric conductive wires, Cu has been widely employed as metallic element in THz and microwave MMs.
- Pt is also a noble metal and precious in the market. Apart from the usage in jewelry, Pt keeps the unique position in catalysis because of its low overpotential ($\sim$30 mV) for H_2 evolution reaction. Although there have been numerous trials to replace Pt with other compounds, the efficiency of Pt in the catalysis reaction has not been defeated to date (Vesborg et al., 2015). Also, Pt is used for fuel cells and will be more valuable in the industrial aspect.
- Ti is a well known transition metal and abundant in nature, exhibiting corrosion resistance like Au and Pt, due to the surface oxidation. The relative permittivity in Fig. 1.6c indicates that the imaginary part is smaller than Cu, Pt, and Ni. Thin Ti layer can be incorporated in plasmonic resonators to increase adhesion of Au to transparent insulators like SiO_2. Practically, Ti is mostly used in alloys and a component in cemented carbides. Also, it is quite well known that Ti–Ni alloy is a shape-memory alloy.
- Ni belongs to the series of iron in the atomic periodic table. The relative permittivity in Fig. 1.6d is qualitatively similar to that of Pt and is consistent with the white glossy appearance. Ni is basically stable, which is relatively less oxidized than Fe and is quite stable for alkali. Therefore, Ni has various practical usages such as plating. It is also known that Ni alloys are practical as magnetic and heat-resistance materials. Furthermore, Ni has an advantage in the absorption of H_2 and N_2 gases.

Besides, we note that Ti and Ni are frequently used as mask for dry etching in the current nanofabrication process.

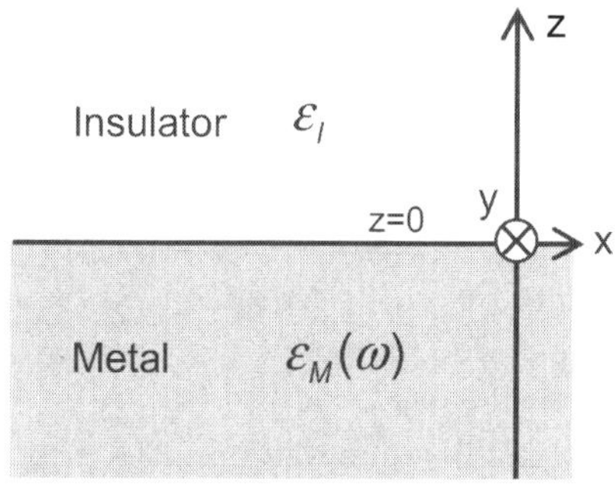

Figure 1.7 Single MI interface between semi-infinitely thick M and I layers. The MI interface is set to the xy plane at $z = 0$.

1.3 Metal–Insulator Interface Where Surface Plasmon Polaritons (SPPs) Emerge

SPP exists at single metal-insulator (MI) interface of semi-infinitely thick metal and insulator as shown in Fig. 1.7. The SPP is a surface-wave solution of Maxwell equations. In this section, the dispersion equation (DE) of the SPP is explicitly derived and the basic properties are described.

Localized waves at the interface are expressed as exponentially decreasing functions in each semi-infinite region. In the spatial coordinate of Fig. 1.7, the component propagating along the z axis has a factor of $\exp(ik_z z)$ in Eqs. 1.13 and 1.14, and has a constraint such that

$$|\exp(ik_z z)| = \exp(-\mathrm{Im}(k_z)z) \to 0 \text{ as } z \to \pm\infty.$$

Hence, there is a constraint such that

$$\mathrm{Im}(k_z)\begin{cases} > 0 & \text{for } z > 0 \\ < 0 & \text{for } z < 0. \end{cases} \tag{1.43}$$

Boundary conditions under p polarization Before proceeding to derivations of DEs, boundary conditions at interface are explicitly written down. We consider p polarization[b] for SPP. Let us set the

[b] p polarization is sometimes called as transverse-magnetic (TM) polarization, which represents the magnetic-field vector parallel to the interface. The TM polarization is a term in waveguide theory.

plane of incidence to be the xz plane; then the E-field vector $\mathbf{E}$ is parallel to the xz plane (i.e., $\mathbf{E} \parallel xz$ plane) and the magnetic field vector $\mathbf{H}$ is parallel to the y axis (i.e., $\mathbf{H} \parallel y$). The boundary conditions connect the components of E_x, H_y, and $(\varepsilon E)_z$. Note that $E_y = 0$, $H_x = 0$, and $H_z = 0$.

Equation 1.7 is non-zero only for the y component and is rewritten as

$$\partial_z E_x(\mathbf{r}) - \partial_x E_z(\mathbf{r}) = i\omega\mu_0 H_y(\mathbf{r}). \tag{1.44}$$

Equation 1.8 is non-zero for x and z components. The x component is explicitly

$$-\partial_z H_y(\mathbf{r}) = -i\omega\varepsilon_0\varepsilon(\omega)E_x(\mathbf{r}). \tag{1.45}$$

Another Eq. 1.9 is rewritten as

$$\varepsilon(\omega)\{\partial_x E_x(\mathbf{r}) + \partial_z E_z(\mathbf{r})\} = 0. \tag{1.46}$$

Let us remind that we are here considering localized waves at the MI interface. In this situation, the harmonic solutions of Eqs. 1.44–1.46 have z-dependence such as

$$F(\mathbf{r}) = F(x)\exp(ik_z z) \tag{1.47}$$

where $F = E_i$ or $F = H_j$, and k_z obeys the constraint in Eq. 1.43.

The E_z component is expressed as a function of E_x by use of Eqs. 1.46 and 1.47:

$$E_z(x) = -\frac{\partial_x E_x(x)}{ik_z}. \tag{1.48}$$

Similarly, using Eqs. 1.45 and 1.47, the H_y component is explicitly written by E_x:

$$H_y(x) = \frac{i\omega\varepsilon_0\varepsilon(\omega)}{ik_z}E_x(x). \tag{1.49}$$

Concerning boundary conditions, Eqs. 1.48 and 1.49 are pivotal because they help to lead a wave equation represented with E_x by eliminating E_z and H_y.

A little further modification of equations leads to the equation for E_x. By substituting Eqs. 1.48 and 1.49 into Eq. 1.44, we obtain the wave equation in coordinate representation:

$$ik_z E_x(x) + \frac{1}{ik_z}\partial_x^2 E_x(x) = -\frac{\omega^2\varepsilon_0\mu_0\varepsilon(\omega)}{ik_z}$$

$$\therefore\ \partial_x^2 E_x(x) = -\left\{\left(\frac{\omega}{c_0}\right)^2 \varepsilon(\omega) - k_z^2\right\}E_x(x) \tag{1.50}$$

where the well-known relation of $\varepsilon_0\mu_0 = c_0^{-2}$ was used. We here define K_x such as

$$K_x = \sqrt{\left(\frac{\omega}{c_0}\right)^2 \varepsilon(\omega) - k_z^2}. \tag{1.51}$$

The K_x depends on k_z and ω, and includes the relative permittivity ε. The general solution of Eq. 1.50 is expressed as

$$E_x(x) = A\exp(+iK_x x) + B\exp(-iK_x x) \tag{1.52}$$

where A and B are constants. Thus the K_x has physical meaning as the x component of wavevector. The sign $\pm$ in Eq. 1.52 represents propagating directions for the x axis.

Dispersion diagram of SPP at single MI interface In the $z > 0$ region of insulator in Fig. 1.7, the x-component of electric field is explicitly written as

$$E_x^{(\mathrm{I})}(x, y, z) = E_x^{(\mathrm{I})}(x)\exp\left(ik_z^{(\mathrm{I})}z\right) \tag{1.53}$$

where $\mathrm{Im}(k_z^{(\mathrm{I})}) > 0$. Similarly, in the $z < 0$ region of metal,

$$E_x^{(\mathrm{M})}(x, y, z) = E_x^{(\mathrm{M})}(x)\exp\left(ik_z^{(\mathrm{M})}z\right) \tag{1.54}$$

and $\mathrm{Im}(k_z^{(\mathrm{M})}) < 0$. Note that the plane of incidence is here set to be the xz plane, so that $E_x^{(\mathrm{I})}$ and $E_x^{(\mathrm{M})}$ are independent of y in the right-hand side of Eqs. 1.53 and 1.54.

One of the Maxwell boundary conditions is $H_y^{(\mathrm{I})}(x) = H_y^{(\mathrm{M})}(x)$ at $z = 0$. By use of Eq. 1.49, the following Eq. 1.55 holds:

$$\frac{i\omega\varepsilon_0\varepsilon_I}{ik_z^{(\mathrm{I})}}E_x^{(\mathrm{I})}(x)|_{z=0} = \frac{i\omega\varepsilon_0\varepsilon_M}{ik_z^{(\mathrm{M})}}E_x^{(\mathrm{M})}(x)|_{z=0}. \tag{1.55}$$

The x-components also satisfy $E_x^{(\mathrm{I})}(x)|_{z=0} = E_x^{(\mathrm{M})}(x)|_{z=0}$. Thus one reaches the following relation represented with ε and k_z:

$$\frac{\varepsilon_I}{k_z^{(\mathrm{I})}} = \frac{\varepsilon_M}{k_z^{(\mathrm{M})}}. \tag{1.56}$$

Let us remind the definition of K_x (Eq. 1.51) and the continuous condition $K_x^{(\mathrm{I})} = K_x^{(\mathrm{M})}$ at $z = 0$. Therefore we simply express the x-component as $K_x[= K_x^{(\mathrm{I})} = K_x^{(\mathrm{M})}]$ and rewrite Eq. 1.56 with K_x:

$$\frac{\varepsilon_I}{\sqrt{\left(\frac{\omega}{c_0}\right)^2 \varepsilon_I - K_x^2}} = \frac{\varepsilon_M}{-\sqrt{\left(\frac{\omega}{c_0}\right)^2 \varepsilon_M - K_x^2}}. \tag{1.57}$$

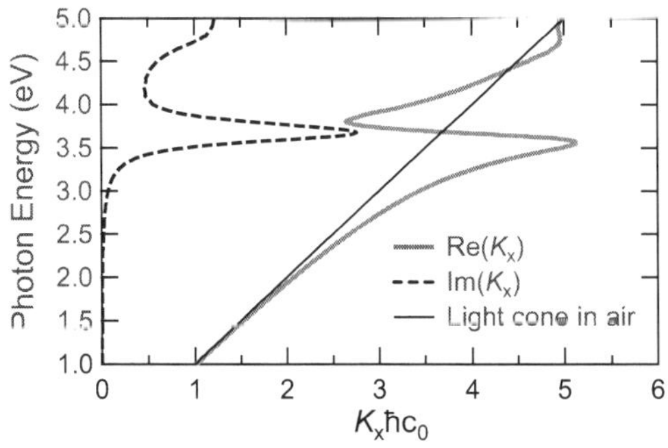

Figure 1.8 Dispersion diagram at the actual MI (Ag–air) interface, evaluated from Eq. 1.58 and plotted on $(K_x\hbar c_0, \hbar\omega)$ plane. $\mathrm{Re}(K_x)$ is shown with thick solid line, $\mathrm{Im}(K_x)$ with dashed line, and light cone in air with thin oblique line. Permittivity of Ag is shown in Fig. 1.5a. Permittivity of air was set to 1.00054.

Note that the sign of k_z in Eq. 1.56 was determined using Eq. 1.43. Equation 1.57 is DE at the interface of $z = 0$, determines the ω-dependence of K_x, and represents basic properties of the interface waves such as SPPs. To transform the SPP dispersion into a usual form, a little efforts to modify Eq. 1.57 lead the following:

$$\frac{\omega}{c_0}\sqrt{\frac{\varepsilon_M(\omega)\varepsilon_I}{\varepsilon_M(\omega)+\varepsilon_I}} = K_x. \tag{1.58}$$

The left-hand side in Eq. 1.58 has only ω-dependent terms and the right-hand side comprises K_x.

Figure 1.8 shows a realistic dispersion at Ag–air interface, evaluated by using Eq. 1.58. Permittivity of Ag was taken from Fig. 1.5a and that of air was set to 1.00054. $\mathrm{Re}(K_x)$ and $\mathrm{Im}(K_x)$ are shown with thick solid and black dashed lines, respectively. Note that the dispersion is plotted on $(K_x\hbar c_0, \hbar\omega)$ plane using the $\hbar c_0$-multiplied Eq. 1.58. Qualitatively, there are three regions as follows.

(i) At the low energy range below 2 eV, the SPP dispersion is almost real-valued, and is very close to and located outside the light cone in air (thin oblique line).
(ii) At the energy range from 2.0 to 3.8 eV, the dispersion deviates from the light cone and the $\mathrm{Im}(K_x)$ increases due to the *d–d* interband transition. It is interesting to point out that the group

velocity v_g of the interface wave, defined by

$$v_g = \frac{\partial \omega}{\partial K_x}, \tag{1.59}$$

becomes negative at that range from 3.6 to 3.8 eV. These dispersion behaviors at flat Ag interface were experimentally observed in 1970s (Arakawa et al., 1973; Alexander et al., 1974). In this energy range from 3.6 to 3.8 eV, backward surface wave propagating for $-K_x$ direction is excited while incident light has $+K_x$ wavenumber.

(iii) Above 3.8 eV, both interband transitions and Drude-like permittivity contribute to the dispersion. The imaginary part $\mathrm{Im}(K_x)$ has a peak at 3.8 eV, implying that the interband transition noted in Section 1.2.3 has large oscillator strength. Consequently, it is not true to simply call it *SPP* dispersion in this range. In fact, the interface wave is excited by far-field at 3.8 eV because the dispersion is located inside the light cone. This means that, at 3.8 eV, the wave induced at the interface is an exponentially decreasing wave rather than a surface propagating wave. The validity of Drude-metal approximation for actual Ag is limited to the energies below 3 eV.

On analytical solution More analytical expression of Eq. 1.58 has been derived in some textbooks (for example, Sakoda, 2005) by assuming the real-valued Drude permittivity such that

$$\varepsilon_M(\omega) = 1 - \left(\frac{\omega_p}{\omega}\right)^2. \tag{1.60}$$

For simplicity, let us assume here $\varepsilon_I = 1$ and then solve Eq. 1.58 for ω.

$$\frac{\omega^4}{\omega_p^2 c_0^2} - \left(\frac{1}{c_0^2} + \frac{2K_x^2}{\omega_p^2}\right)\omega^2 + K_x^2 = 0$$

$$\therefore \omega^2 = \frac{\omega_p^2 + 2(K_x c_0)^2}{2} \pm \frac{\sqrt{\omega_p^4 + 4(K_x c_0)^4}}{2}.$$

We finally reach

$$\hbar\omega = \sqrt{\frac{(\hbar\omega_p)^2 + 2(K_x \hbar c_0)^2 \pm \sqrt{(\hbar\omega_p)^4 + 4(K_x \hbar c_0)^4}}{2}}, \tag{1.61}$$

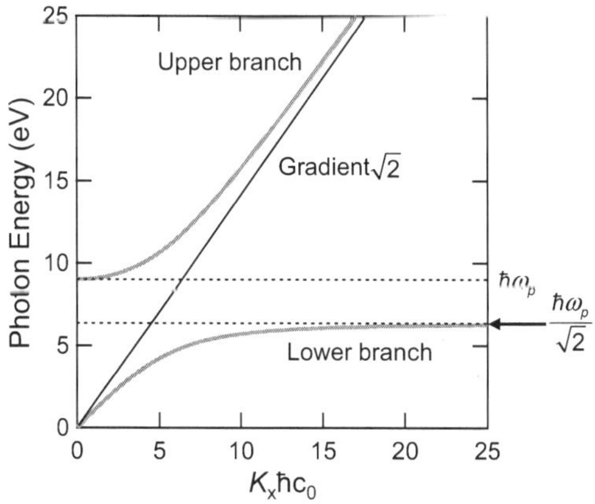

Figure 1.9 Disperion diagram of purely theoretical Drude-metal–air interface (thick solid curves), obtained from analytic Eq. 1.61. Dotted lines indicate the energies of $\hbar\omega_p$ and $\hbar\omega_p/\sqrt{2}$ where $\hbar\omega_p = 9.01$ eV. The plasma frequency ω_p of Ag was taken from Table 1.1.

by multiplying $\hbar$ for the plot in Fig. 1.9. The lowest branch, that is, SPP dispersion, corresponds to the minus of $\pm$ inside the root square in Eq. 1.61. On the other hand, the plus of $\pm$ corresponds to the upper branch. Note that the permittivity is assumed to be purely real; therefore the dispersion relation is also real-valued.

Figure 1.9 shows dispersions (thick solid curves) obtained from Eq. 1.61. The bulk plasma energy $\hbar\omega_p$ was set to 9.01 eV by using the ω_p of Ag (see Table 1.1) for comparison with Fig. 1.8. The lower branch represents SPP, going asymptotically to $\hbar\omega_p/\sqrt{2}$ as $K_x \to \infty$. It is evident that the dispersion is very different from the dispersion on actual Ag in Fig. 1.8. The vertical axis in Fig. 1.8 is plotted for the photon-energy range of (1, 5) eV while the vertical axis in Fig. 1.9 is plotted for (0, 25) eV.

On the other hand, in Fig. 1.9, the upper branch appears for positive ε_M, which is not surface wave but is a transverse wave. The upper branch appears only inside the light cone and is excited with far-field; besides, the upper branch asymptotically becomes close to the line of gradient $\sqrt{2}$, that is, $\omega/c_0 = \sqrt{2}K_x$. We note again that the actual permittivity in metal cannot be approximated only by the

Drude permittivity due to the interband transitions; therefore, the upper-branch wave is not observed actually.

1.4 Brief Overview of the History

In this section, we briefly outline the history related to plasmonic resonators. Concretely, SPPs, extraordinary transmission, and MMs are addressed.

1.4.1 *Early Studies on SPPs*

As noted in Section 1.1, bulk plasmon was first considered in ionized gas (i.e., plasma) (Tonks and Langmuir, 1929), describing the longitudinal collective oscillation at the plasma frequency ω_p. Good conductive metals are approximately understood to have free electrons inside them. However, it is difficult to probe inside the bulk metals; we can practically probe the surface with EM waves from microwaves to UV light. Thus, the plasmonic resonances have been usually probed under the surface illuminations. Nowadays, it has been well known that surface of bulk metals, thin films of metals, and nanostructures of metals show quite different spectral behaviors, and they are key subjects in this book.

The surface plasmon studies in the early stage from 1970s to 1980s are listed as follows (for example, (Raether, 1988)).

(i) SPPs on smooth and single metal/dielectric interface.
The dielectric includes air and vacuum. This configuration has been already addressed in Section 1.3 and the SPP dispersion was derived (Figs. 1.8 and 1.9). Ideal (or atomic-level) smoothness of the interface was assumed.

(ii) SPPs on thin metallic film.
Single crystals of metals need particular preparation in the growth and polishing. Instead, thin films of metals are and have been used so often.
Thin metallic films deposited on a bulk SiO_2 or a total reflection prism exhibit SPP-origin resonances, which are easily detected in incident angle resolved experiment. The

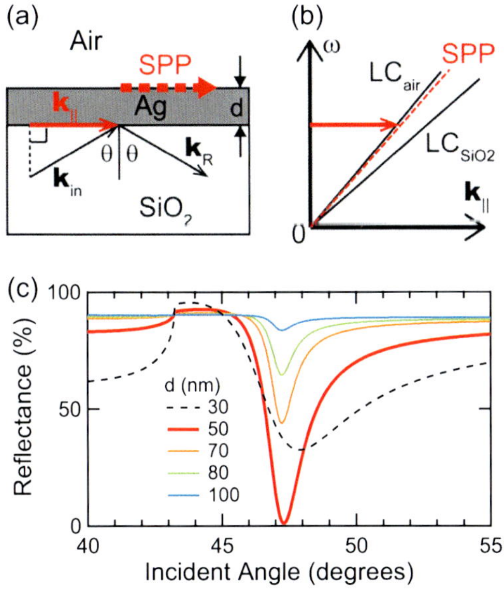

Figure 1.10 (a) Total reflection configuration for excitation of SPP on a thin Ag film. (b) Dispersion diagram of light cones (LCs, oblique solid lines) and SPP at air side (red dashed line). (c) Incident-angle-resolved R spectra for various film thickness d. Incident wavelength was 500 nm. The R spectra were numerically calculated.

configuration is schematically drawn in Fig. 1.10a; incidence comes from the side of SiO_2 substrate with incident angle θ and sheds on a thin Ag film of thickness d. In this configuration, SPP can be excited at the air side, as schematically drawn with a red bold dashed-line arrow. The configuration is sometimes referred to as Kretschmann configuration; besides, it is often called attenuated total reflection (ATR) configuration.

Figure 1.10b is a schematic of dispersion diagrams in the ATR configuration. The diagrams are represented in the plane of angular frequency ω and wavenumber component parallel to the flat interface $k_{\parallel}$. Oblique solid lines denote light cones (LCs) in air and SiO_2. LC is generally expressed as

$$\omega = \frac{c_0}{n} k_{\parallel} \tag{1.62}$$

where n is refractive index of media. Note that LC_{air} appears at relatively upper side than LC_{SiO2}, due to the smaller refractive index than that in SiO_2. SPP at the air side, shown with red dashed line, is located just below LC_{air}, as described in Section 1.3. The diagrams suggest that incidence from the SiO_2 side can excite the SPP at the air side when the near-field induced at the Ag/SiO_2 interface is able to reach the interface of air/Ag through the thin metallic film.

Figure 1.10c shows numerically calculated R spectra dependent on incident angle θ. The incident wavelength was set to a visible wavelength of 500 nm (i.e., 2.479 eV in photon energy); the realistic SPP dispersion of air/Ag interface is seen in Fig. 1.8. Varying the thickness d, deep R dip that reaches almost 0% appeared at $d = 50$ nm and $\theta = 47.3^\circ$ (red solid curve); the dip corresponds to excitation of the SPP. As the thickness d increases, the R dip gradually becomes small and almost disappears at $d = 100$ nm (colored thin solid curves). This behavior comes from separation of air/Ag and Ag/SiO_2 interfaces by the thick Ag film. At $d < 50$ nm, the thin films also allows far-field transmission through the film and the R dips exhibit broadening (black dashed curve), implying the weak confinement of SPP at the air/Ag interface and larger radiation loss. The kinks of R spectra at $\theta = 43.25^\circ$ indicate the critical angle for total reflection at the air/Ag interface. The R spectra were calculated using scattering matrix algorithm described later (Section 1.5.1) and measured permittivity of Ag (Rakić et al., 1998).

SPP on thin metallic film is currently implemented in surface plasmon resonance (SPR) sensors because the resonance is highly sensitive to the incident angle as shown in Fig. 1.10c. In actual experiment, the angle is resolved at high resolution less than 0.1°. Thus, slight change of refractive index at the air/Ag interface can be detected at high sensitivity. The SPR sensor configuration is shown in Fig. 1.11.

(iii) SPPs on rough surfaces.

When considering metallic film making techniques using vacuum deposition, the interface has often roughness due to the grains of metal. Roughness can also be introduced by artificial treatments, enabling external far-field to couple with the SPPs without any coupler such as total-reflection prism.

(iv) Local plasmons in metallic nanoparticles.
When metal deposition is ceased at a short time (~10 s), metallic nanoparticles (or islands) are left on the substrate. Thus, metallic islands were feasibly accessible and studied by many researchers in parallel to the thin metallic films noted above.

Au nanoparticles have a far long history though they were not studied extensively in 1980s; they were chemically synthesized and the colloidal solutions were made in the 19th century (Faraday, 1857) though they did not attract much interest at the time. Empirical production of metallic nanoparticles was made far long time ago; indeed, a colored antique, so-called The Lycurgus Cup, in the ancient Roman period includes Au–Ag alloy nanoparticles, which is preserved at the British Museum.[c] Scattered light by metallic nanoparticles exhibits various colors in accordance with the size; the physical properties were resolved at the beginning of the 20th century (Maxwell-Garnett, 1904, 1906). Besides, shapes also contribute to the scattering spectra (see Section 3.2). Stimulated by the SPP studies, metallic nanoparticles were revived and studied by many researchers since 1990s.

Figure 1.11a schematically illustrates local plasmon in a Au nanoparticle, whose surface was modified by adding "capture" molecules (Y-like shapes) for analytes (i.e., target molecules). The capture of the target is usually recognized in extinction spectra that are evaluated from transmission spectra of the nanoparticle-including solution; the resonant shift was observed in a similar manner to the SPR sensors mentioned above (for example, Anker et al., 2008). Also, resonant E fields on the local-plasmon resonance are expected to enhance molecular vibration signals. The local plasmon is confined in the nanoparticle and therefore needs to capture the analytes at small distance δ_d or almost on the immobilized surface. We

[c]http://www.thebritishmuseum.ac.uk and search "Lycurgus Cup."

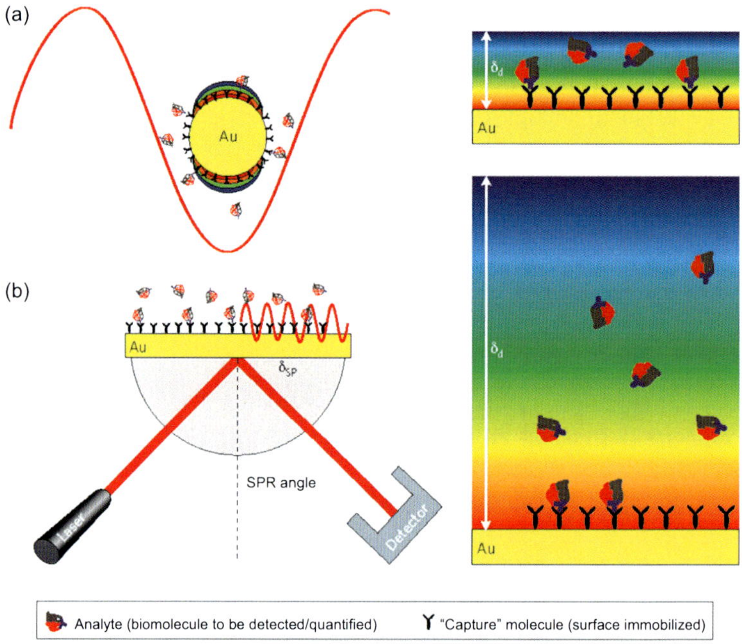

Figure 1.11 Schematics based on plasmonic resonances. (a) Local plasmon in a Au nanoparticle (or nanosphere) and the sensing configuration. The inset at the right-hand side enlarges near the surface. (b) SPP on thin Au film deposited on a total reflection prism, represented by red sinusoidal curve, and SPR sensing configuration. Reprinted with permission from Macmillan Publishers (Brolo, 2012), copyright 2012.

note that the surface engineering to use the capture molecules is the technique available widely after 2000 (not in 1980s).

Figure 1.11b shows SPP on a thin Au film. Total reflection prism enables the SPP excitation, as described in Fig. 1.10. In comparison with the local plasmon, the SPP on the thin film is able to detect the analytes at larger distance δ_d.

(v) SPPs on shallow gratings.
Sinusoidal-profile gratings were available in 1980s. By depositing metals such as Ag and Al on the gratings, SPPs were excited with far-field illumination. In contrast to (ii) above, the well-defined periodicity makes it possible to observe dispersion diagrams of SPPs, which are reduced into the first

Brillouin zone. The depth profiles of the grating were limited to shallow cases such as depth $\ll$ periodicity from both experimental and theoretical aspects. Numerical methods for metal-included systems were established in 1990s, as noted later (see Section 1.5).

(vi) Second harmonic generation (SHG) on SPP resonances.
SHG from smooth metallic interfaces was examined experimentally and theoretically (Bloembergen et al., 1968). SHGs from rough metallic surfaces were also studied in 1980s (Chen et al., 1981; Boyd et al., 1984). Surface roughness on Ag film enhanced SHG by 10^4 times in comparison with flat Ag surface and besides resulted in broad light emission around the SHG (Chen et al., 1981); the surface treatment using KCl was likely to affect Ag on the surface and to yield chemical species different from pure Ag. It was also suggested that Ag nanostructures are quite likely to contribute to SHG.

(vii) Surface-enhanced Raman scattering (SERS) on SPP resonances.
SERS was experimentally found in 1974 (Fleischmann et al., 1974). Plasmonic enhancement was expected and kept being explored in 1980s though the attempts at the time were not so successful. The recent successful development is described later (see Section 5.6.1).

(viii) Tunneling of electrons in metal-insulator-metal (MIM) configuration.
From analogy from Josephson coupling using superconductors in 1962 (Josephson, 1962), MIM and metal-insulator-semiconductor (MIS) structures were expected to serve as tunnel junction. The theoretical studies on the configurations were already reported in 1960s, and clarified the waveguide and localized modes in the structures (Swihart, 1961; Economou, 1969).

Tunneling-current-induced light emission was experimentally observed in flat MIM and MIS junctions with low efficiency $\sim 10^{-4}$; by adding metallic nanoparticles on the flat surface of a buried insulator layer in the MIM or MIS structures, it was found that the light emission is more than 10 times enhanced (Lambe and McCarthy, 1976; McCarthy

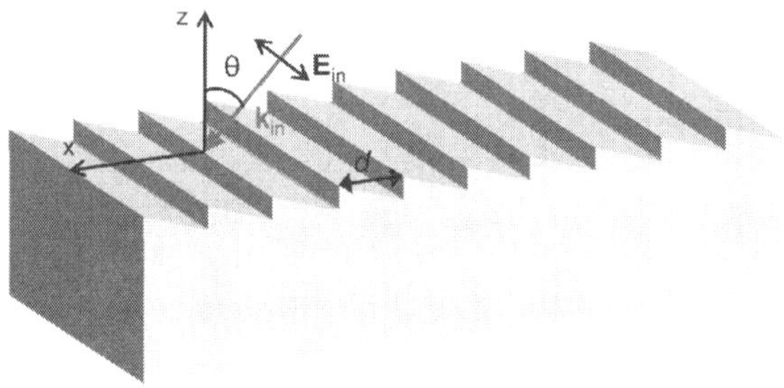

Figure 1.12 Schematic of a diffraction grating, composed of 1D periodic structure of sawtooth shape, called echelette grating. The surface is usually coated by a metallic thin film. The periodicity is comparable to working wavelengths.

and Lambe, 1977, 1978). The metallic nanoparticles are considered to work as effective light scatters to the outer domain.

Diffraction Gratings Figure 1.12 shows a schematic of diffraction grating, which is one-dimensional (1D) periodic and called echelette grating with sawtooth shapes. The surface is usually coated by a metallic thin film to obtain high reflective responses. Diffraction gratings are used in monochromators and one of the well-established optical elements. When the coordinate axes are set in Fig. 1.12, SPP on the grating is excited with p polarization (i.e., $\mathbf{E}_{\rm in} \| xz$ plane).

Figure 1.13 shows a typical grating and the optical responses. We here refer to lamella grating as an example. The structure is drawn in Fig. 1.13a and consists of periodic array of metallic (Ag) rods; the grating is periodic along the x axis and is assumed to be infinitely long along the y axis. The periodicity a, slit width w, and thickness t are the structural parameters. The grating exhibits prominent optical responses for p polarization, that is, $\mathbf{E}_{\rm in} \| xz$ plane where $\mathbf{E}_{\rm in}$ denotes incident polarization.

Figure 1.13b displays optical spectra of R, transmittance (T), and the diffractive components (R_{mn} and T_{mn}) at normal incidence with $\mathbf{E}_{\rm in} \| x$. The spectra were numerically calculated using RCWA+S-matrix method described in Section 1.5.1. The structural parameter

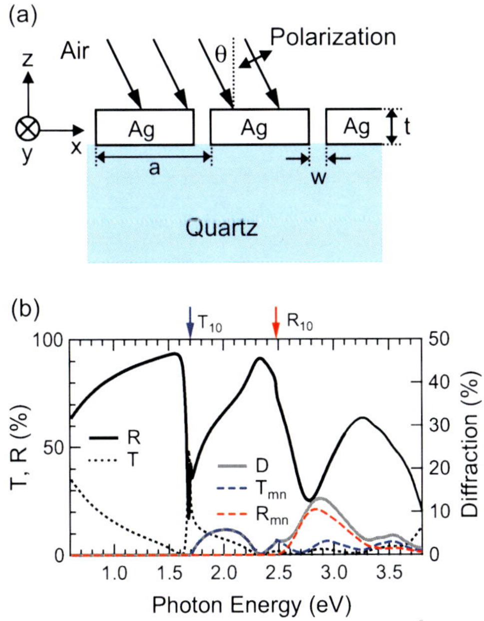

Figure 1.13 Typical optical responses by a grating. Lamella grating is taken as an example. (a) Schematic of the structure, which comprises periodic array of metallic rods. (b) Optical spectra for the lamella grating of a = 500 nm, w = 100 nm, and t = 100 nm. R and T spectra at the normal incidence ($\theta = 0°$) with x polarization are shown with black solid and dotted lines, respectively. Total diffraction (D) spectrum is plotted for the right axis, shown with gray line. Sums of transmissive and reflective diffraction components, T_{mn} and R_{mn}, are shown with blue and red dashed lines, respectively. Blue and red arrows indicate the energies of $T_{\pm1,0}$ and $R_{\pm1,0}$ channels opened, respectively.

was set to (a,w,t) = (500,100,100) nm. The main properties are seen as deep dips in the R spectrum (black solid line). The R dips at 1.7 and 2.8 eV correspond to diffraction channels. At 1.7 eV indicated by blue arrow, the lowest diffraction channels $T_{\pm1,0}$ become open, which are transmissive components; the sum is written as T_{mn} (blue dashed line). At 2.48 eV, both transmissive and reflective channels are open. The sum of reflective diffraction is written as R_{mn}. The total diffraction (D, gray line), defined by $D = T_{mn} + R_{mn}$, takes the maximum at 2.8 eV.

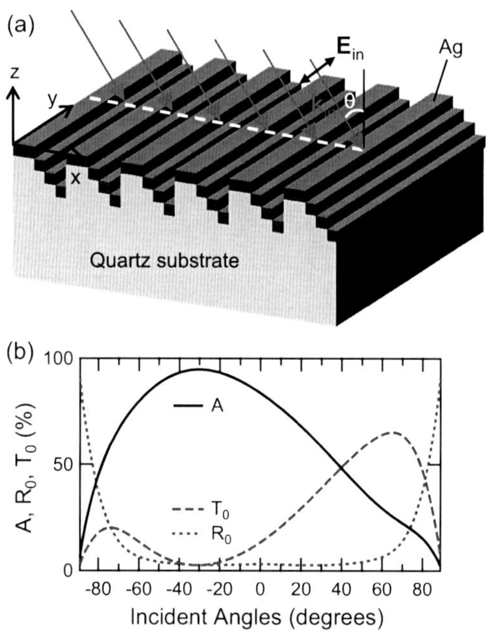

Figure 1.14 A deeply asymmetric light absorber, which is 1D periodic and has step-like structure. (a) Schematic of structure, which has periodicity of 250 nm and comprises three steps. The top Ag rod has the *xz* section of 100×50 nm^2 and other rods have 50×50 nm^2 sections. (b) Numerically calculated absorbance (A) spectrum at 620 nm that depends on incident angle θ is shown with black line. The zeroth-order reflectance (R_0, dotted line) and the zeroth-order transmittance (T_0, dashed line) spectra are also shown. Adapted from Iwanaga (2012a).

Deeply asymmetric light absorber Figure 1.14 shows a deeply asymmetric light absorber, which mainly serves not as a diffraction grating but as a subwavelength structure free from diffraction. The structure is 1D periodic along the *x* axis, as shown in Fig. 1.14a. The unit cell comprises a step-like structure of quartz with 250 nm periodicity and has 50 nm thick Ag on the top. The depth of each step was also set to be 50 nm. The width of the on-top Ag structure was 100 nm and other widths of Ag were 50 nm. The plane of incidence is set to the *xz* plane. Let us consider the case that p-polarized incidence (i.e., $\mathbf{E}_{in} \| xz$ plane) travels from the top side of air with angle θ.

Figure 1.14b presents numerically calculated light absorbance (A), the zeroth-order transmittance T_0, and the zeroth-order reflectance R_0 at the incident wavelength of 620 nm (i.e., 2.0 eV in photon energy). Note that the zeroth-order components are ordinary T and R. The A was defined in % by

$$A = 100 - \sum_{n=0,\pm 1,\pm 2,\cdots} (T_n + R_n), \tag{1.63}$$

representing linear light absorption. Obviously, the A spectrum is deeply asymmetric for θ and nearly perfect absorption takes place from $\theta = 20^\circ$ to 40°. The polarity of angle θ is defined in accordance with the sign of $\sin\theta$; then, the projection of incident wavevector onto the x axis, that is, $|\mathbf{k}_{\text{in}}|\sin\theta$ has positive (or negative) sign for the $+x$ (or $-x$) direction. We mention that the R_0 spectrum is exactly symmetric for θ due to the reciprocity in this optical configuration. As for the optical reciprocity, a review was reported in (Potton, 2004) and an intuitive proof was reported in (Iwanaga et al., 2007).

The large A in Fig. 1.14 primarily comes from the structures in the unit cell and is distinct from diffraction-associated SPPs that are usually induced in shallow gratings. The diffraction-associated SPPs are sensitive to incident angle whereas the large A appears at wide ranges of incident angles and wavelengths. Besides, the A spectrum at $\theta = 40^\circ$ exceeds more than 90% at 600–800 nm, realizing a nearly perfect absorber, is larger than 75% at 600–1000 nm, and is larger than 50% at 400–1500 nm (Iwanaga, 2012a). Thus, the asymmetric absorber shows broadband responses for wavelengths and incident angles.

1.4.2 *Era of Nanotechnology and Dawn of Plasmonics*

Around 2000, at the beginning of the era of nanotechnology, metallic nanostructures became hot research targets. Perforated metallic films, nanoparticles such as spheres, rods, tetrapods, and so on, were extensively investigated. These subjects are often summarized in a field, called plasmonics. The development of fabrication techniques yielded various nanostructures and have greatly stimulated nanotechnology.

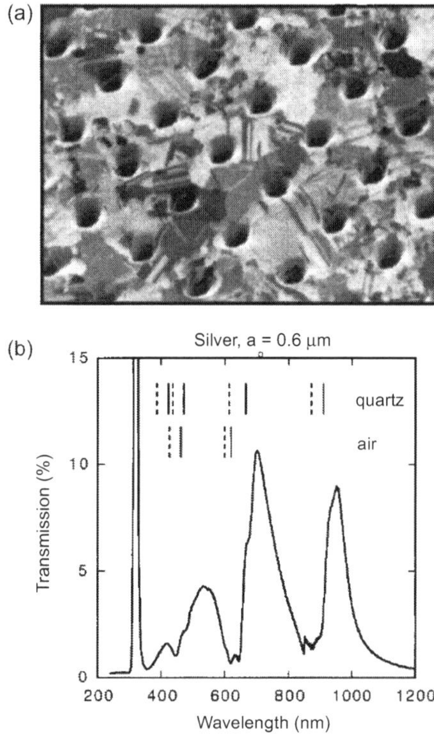

Figure 1.15 A representative result of EOT. (a) A typical SEM image of perforated Ag film. Periodicity was 900 nm, hole diameter 150 nm, and thickness of Ag film was 200 nm. (b) Measured T spectrum at the normal incidence in the case of the periodicity 600 nm. Reprinted with permission from Ghaemi et al. (1998). Copyright 1998 by the American Physical Society.

Extraordinary transmission (EOT) A strong trend related to plasmons was triggered in 1998 by a report on EOT in perforated metallic films (Ebbesen et al., 1998; Ghaemi et al., 1998). The EOT was observed in simple configuration but it was not at all expected and therefore left a great impact on plasmon studies, opening a new era of plasmonics.

Figure 1.15 summarizes the main result on the EOT (Ghaemi et al., 1998). Figure 1.15a shows scanning-electron-microscope (SEM) image of perforated Ag film. The thickness was 200 nm and thick enough for the skin depth of Ag. The periodicity of air-hole array was 900 nm and the hole diameter was 150 nm. The holes were fabricated by a focused-ion-beam (FIB) instrument, which is one of the representative instruments in nanotechnology. Also, SEM is a standard tool to observe nanostructures at nm precision.

Figure 1.15b shows an EOT peak at 980 nm appears at low-energy positions where the diffraction channels (vertical bars) open. Other T peaks appear in the diffractive range and are not usually called EOT. The EOT peak was approximately 9% while the air-hole aperture was $\pi \cdot 75^2/900^2 \approx 2.2\%$; therefore, the EOT peak was fourfold larger than a simply estimated value from the aperture ratio in the unit cell. Although we have called the T peak at 980 nm EOT peak in accordance with the original paper (Ghaemi et al., 1998), such kinds of T peak through nanoapertures have been observed in many cases and became ordinary effects. The EOT peak is located close to the lowest diffraction mode in the quartz substrate, indicated by a vertical short bar in Fig. 1.15; therefore, it is inferred that the diffraction-associated plasmonic resonance contributed to the EOT peak in an implicit way. We mention that dielectric function of the nanostructure significantly changes around resonances, so that optical properties and constants at resonances become far from electron-density-diluted (i.e., perforated) metallic film (see Chapter 2 for dielectric function).

From the today's viewpoint, EOT phenomena are understood as one of *ordinary* effects on plasmonic resonances. Still, thanks to the finding of EOT, it is no doubt that great interest was turned on metallic nanostructures and plasmonics has become an established subfield in optical sciences. The EOT phenomena greatly stimulated numerous studies for a decade, a part of which was collected in review papers (Genet and Ebbesen, 2007; Garcia-Vidal et al., 2010).

Plasmonics after 2000 Plasmonics, which was originally termed for characterizing confined light guiding using metallic nanostructures (Maier et al., 2001), has extended far beyond the initial notion and has been an established subfield in optical sciences. The EOT effect in Fig. 1.15 is now classified into plasmonics.

Early development was attained through extensive studies on metallic nanoparticles such as spheres, rods, triangles and so on (Mock et al., 2002; Murray and Barnes, 2007). In the metallic

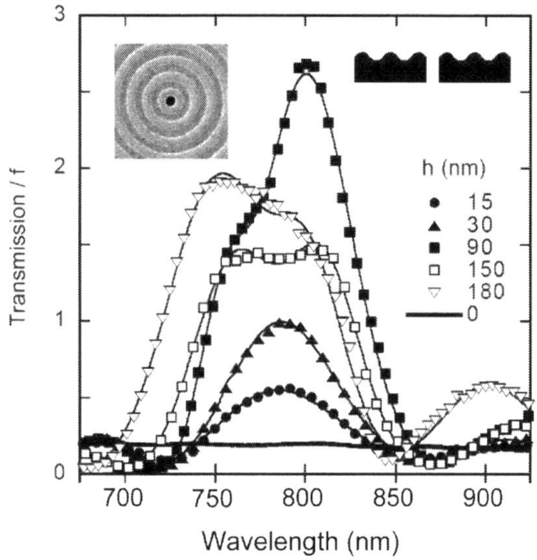

Figure 1.16 Bull's eye structure for EOT. The center of air hole of 440 nm diameter was surrounded by ring structures of sinusoidal profile. The inset shows a top-view SEM image. Transmission spectra are plotted with the various sinusoidal height h. Reprinted with permission from Thio et al. (2001).

nanostructures, Mie resonances are responsible for the origin of the plasmonic resonances (Section 3.2.1).

Figure 1.16 shows a set of measured results on so-called bull's eye, which is an extension of the EOT. The inset shows top-view SEM image of single air hole surrounded by ring structures; the hole was perforated in a free-standing metallic film of Ni 300 nm/Ag 100 nm from the Ni side using FIB. The ring structures were originally introduced on the Ni side. After the perforation, Ag was deposited on the Ni side by 30 nm thickness to coat the surface with Ag.

Figure 1.16 shows transmission spectra normalized by the aperture area. For the air-hole diameter $d = 440$ nm, transmission peaks appeared at 750–800 nm. The diameter of the rings was 1.5, 3.0, 4.5, $\cdots$, 15.0 μm. Varying the height h of sinusoidal structure from 0 to 180 nm, the transmission efficiency was found to be substantially changed. The marks are measured data and thin curves are fitted curves. This result was attributed to surface plasmons on the sinusoidal structures, suggesting that the efficiency to collect light into the subwavelength aperture can be controlled by the structures around the aperture. Indeed, it was reported that photodiode fabricated under the center of Bull's eye show better performance owing to the efficient light collection (Ishi et al., 2005).

Other simple structures such as plasmonic waveguides that comprise MIM structures were also topics in the early stage of plasmonics (Maier and Atwater, 2005) (see Section 3.1 for the plasmonic waveguides).

1.4.3 *Metamaterials (MMs)*

The prefix *Meta* denotes "super" or "beyond". MMs were thus intended to define an upper class of materials. For EM waves, materials are described using permittivity ε and permeability μ; therefore, if one is able to construct artificial structures that have designed values of ε and μ, it means that one has succeeded in *creating* new materials that have not been found in nature. This was initial notion on MMs, which was introduced in the rather optimistic manner. Let us briefly survey the development of MM studies.

Rapid progress in microwave, THz and IR MMs using SRRs The first notion stemmed from a theoretical implication by Pendry et al. for artificial magnetism in a Swiss role structure (Pendry et al., 1999). A key element to exhibit artificial magnetism was introduced. The structure was basically a kind of coils that exhibit electric-field-induced magnetism and was interpreted as a permeability response.

A key experimental demonstration was reported on a microwave MM exhibiting negative refraction, whose appearance is shown in Fig. 1.17a. The Swiss role structure was simplified into double sprit-ring resonator (SRR), which is drawn as the center element in Fig. 1.17b. The details are elaborated in the succeeding heading, "Negative refraction at GHz."

By simply reducing the SRR structure, it was reported that a planar double-SRR arrays exhibit magnetic response around 1 THz (i.e., 300 μm in wavelength) (Yen et al., 2004). The periodicity of the MM was scaled down to 43 μm, so that the fabricated structure was reduced to a planar structure of SRR periodic array with omitting

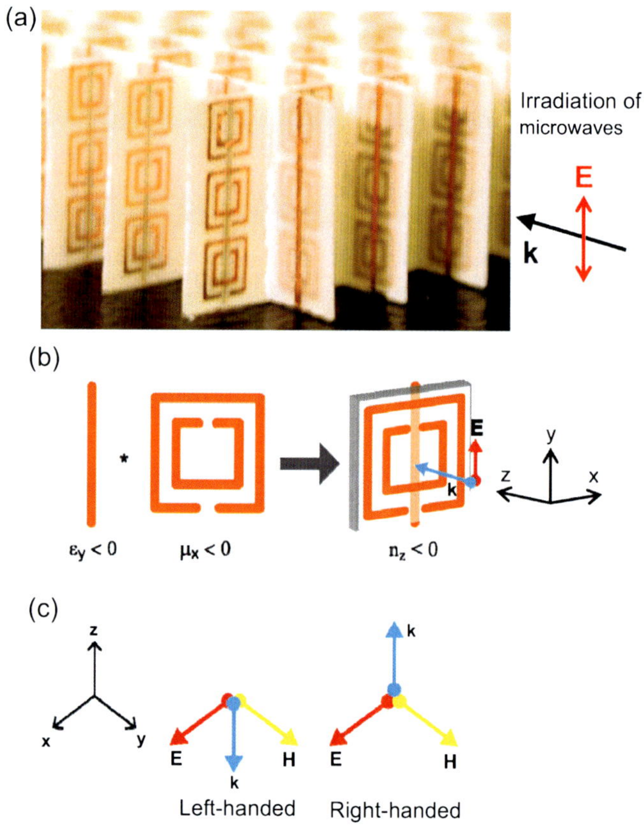

Figure 1.17 (a) Microwave MM of effective negative refractive index. Microwaves were set to irradiate the MM from side. Negative refraction was observed for the vertical polarization. (b) Schematic of the MM unit (or meta-atom) in (a). (c) Definitions of left- and right-handed EM waves. (a) was from Shelby et al. (2001); reprinted with permission from AAAS. (b) and (c) were adapted from Iwanaga (2012c).

the straight wires in Fig. 1.17a. Therefore, only magnetic response was concluded, which was derived from effective permeability extraction procedure described later in this subsection. The metal was Cu because of good conductor at THz and of compatibility in microfabrication technique.

Further reduction of the SRR structure enabled the MM study to make rapid progress over a wide range from microwaves to IR ranges. At IR and NIR ranges, SRRs were more simplified into single SRRs (Linden et al., 2004; Enkrich et al., 2005). It attracted interest where is the limit of the reduction of SRR structures. Several reports consistently concluded that the limit exist at approximately 1 μm periodicity (Ishikawa et al., 2005; Zhou et al., 2005; Klein et al., 2006).

This limit is understood from the permittivity of metals in Figs. 1.5 and 1.6. In a wide frequency range from microwaves to 10^{14} Hz (i.e., 3 μm in wavelength), the real parts of permittivity of Cu, Ag, and Au were approximately constant. Therefore, simple reduction (i.e., scaling) was valid. On the other hand, at optical frequencies more than 10^{14} Hz, analogy does not hold and new designs for negative refraction were needed (Fig. 1.20).

Negative refraction at GHz Extensive MMs studies were triggered by experimental demonstration on negative refraction at microwaves. Figure 1.17a shows a photograph of the MM exhibiting effective negative refraction at microwave frequencies around 10.5 GHz (Shelby et al., 2001). Incidence was shed on the MM in the horizontal plane and the polarization was vertical, parallel to the Cu straight wire, as indicated by arrows. A SRR and straight wire formed a unit cell, and the unit cell was arrayed in a two-dimensional (2D) periodic manner. This experiment provided the first experimental support for the concept of the effectively negative refractive index, being the trigger of extensive MM studies.

Figure 1.17b illustrates the role of the unit cell of the MM in Fig. 1.17a. Effective permittivity ε_y comes from the straight wire, effective permeability μ_x is attributed to SRRs, and the product (Eq. 1.5) is expected to result in effectively negative refractive index n_z. The periodic array in Fig. 1.17a was considered to make the distribution of effective refractive index to be isotropic.

Figure 1.17c shows left-handed and right-handed EM waves. As is well known, ordinary EM waves are right-handed. If EM-wave media have negative refractive index, the EM waves propagate in a left-handed manner. We mention that left-handed EM waves are not limited only to the case of purely real negative refractive index (i.e., $\varepsilon < 0$ and $\mu < 0$).

Effective permittivity and permeability To extract the effective permittivity and permeability in MMs, the so-called retrieval procedure was proposed (Smith et al., 2002).

Figure 1.18a illustrates an optical configuration for a slab of thickness d. When an incident plane wave sheds on the slab at incident angle θ, the complex reflectivity r and complex transmissivity t are related to the wave vector in the slab such that

$$\cos(k_z d) = R(r, t) \tag{1.64}$$

where $R(r, t)$ is a rational expression of r and t (Smith et al., 2002). Figure 1.18b shows a general optical configuration around MM slab

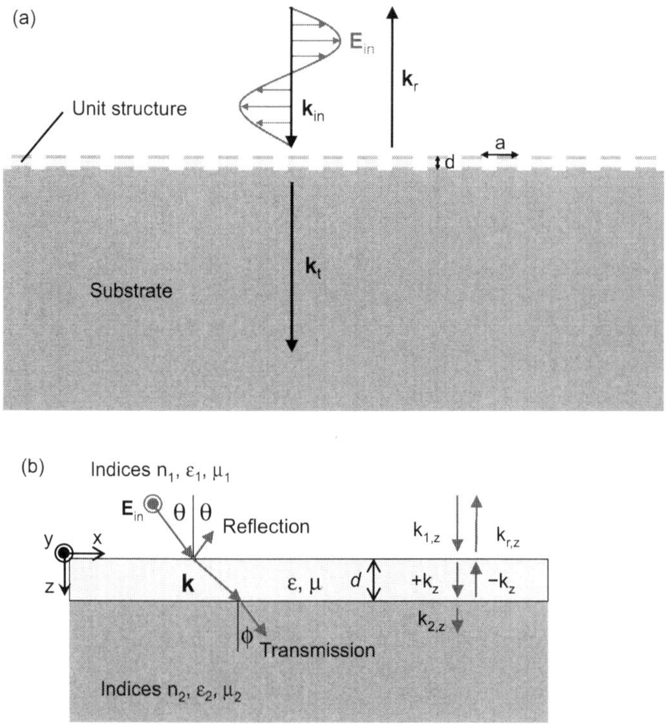

Figure 1.18 (a) Schematic of MMs concept concerning wavelength λ and structural dimensions of periodicity a and thickness d ($d \leq a < \lambda$). Also, a typical optical configuration of incidence, transmission, and reflection is shown. (b) Optical configuration around a homogenized MM slab. Incidence of s polarization is drawn.

under s polarization; then, the concrete expression of the $R(r, t)$ for biaxial MM[d] is written as

$$R(r, t) = \frac{\delta\tilde{t}^2 + \gamma(1 - r^2)}{\tilde{t}\{\gamma(1 + r) + \delta(1 - r)\}} \tag{1.65}$$

where $\tilde{t} = t\exp(ik_2d\cos\phi)$ and ϕ is refraction angle in the substrate. For s polarization,

$$\delta = n_2\cos\phi/\mu_2 \text{ and } \gamma = n_1\cos\phi/\mu_1;$$

for p polarization (i.e., $\mathbf{H}\|y$),

$$\delta = n_2\cos\phi/\varepsilon_2 \text{ and } \gamma = n_1\cos\phi/\varepsilon_1.$$

Note that Eq. 1.65 includes the case of oblique incidence. The z-component of refractive index n_z is related to the k_z as

$$n_z = \frac{c_0}{\omega}k_z$$

at the normal incidence. Therefore the effective refractive index n_z is extracted from

$$n_z = \frac{c_0}{\omega d}\{\arccos[R(r, t)] \pm 2\pi m\} \tag{1.66}$$

where m is a non-negative integer. Since Eq. 1.66 has ambiguity in the real part, it is always necessary to decide how to determine a specific m branch. At a single wavelength measurement or calculation, it is, in principle, difficult to determine which m branch should be taken; at least, some additional examination is necessary. In ordinary ellipsometry for solid materials, a sequence of spectroscopic measurement is conducted to determent the branch. The method may be applicable to MMs.

The succeeding procedure after Eq. 1.66 is as follows.

(i) The admittance, which is n_z/μ_y for s polarization, is uniquely determined. We note that the admittance is independent of Eq. 1.66.
(ii) Once one is able to determine the value of n_z, the y component μ_y is determined.

[d]Generally, isotropic and uniaxial MMs are included in biaxial MMs. Here, the description using Eq. 1.65 does not include MMs of off-diagonal components in the effective ε and μ tensors; therefore, Eq. 1.65 is not directly applicable to chiral MMs.

(iii) Finally the remaining unknown term ε_x is determined from the dispersion equation $\varepsilon_x \mu_y = n_z^2$.

For p polarization, the admittance is n_z/ε_y and the extraction is done in the order of $n_z \to \varepsilon_y \to \mu_x$.

The retrieval procedure assumes that the macroscopic material parameters ε and μ are well defined in the medium, and then provides a procedure to extract ε and μ in a purely algorithmic manner. When the medium is isotropic, the parameters ε and μ are fully determined in the procedure described above. The macroscopic material parameters assume that the long wave approximation for the medium is well-defined. The physical meanings of the retrieval parameters in MMs have been much debated (Koschny et al., 2003; Depine and Lakhtakia, 2004; Efros, 2004; Koschny et al., 2004).

These controversy stimulated reconsiderations of the framework using the macroscopic material parameters. A reliable solution was provided based on microscopic light–matter interaction (Cho, 2010). The formalism is introduced in Section 2.4.

Currently, the extraction of effective (ε, μ) has not been a central issue in the MM studies. This is partially because the justifications are not automatically held for most MMs. Instead, it is gradually becoming a standard to fully clarify the EM eigenmodes (Iwanaga, 2011, 2012b,c; Cho et al., 2014), which are described in Sections 3.6.2 and 3.6.5.

Cloaking Further attracting demonstration on MMs was done as cloaking (Schurig et al., 2006). Figure 1.19 shows outlook of a microwave MM exhibiting cloaking effect. Figure 1.19a presents the photograph of the disk-shaped MM. The wire units shown in the insets (black lines) were loaded on dielectric film of $\varepsilon = 2.33$ at 10 GHz. The units were designed considering effective materials parameters such as μ_r and μ_θ in the 2D polar coordinate (r, θ).

Figure 1.19b shows four E-field patterns: **A** simulated cloaking of Cu cylinder using an ideal parameter set on effective (ε, μ) without any loss, **B** simulated cloaking using a parameter set with loss, realized in the experiment, **C** measured E-field steady state in a configuration that contains the bare Cu cylinder, and **D** measured

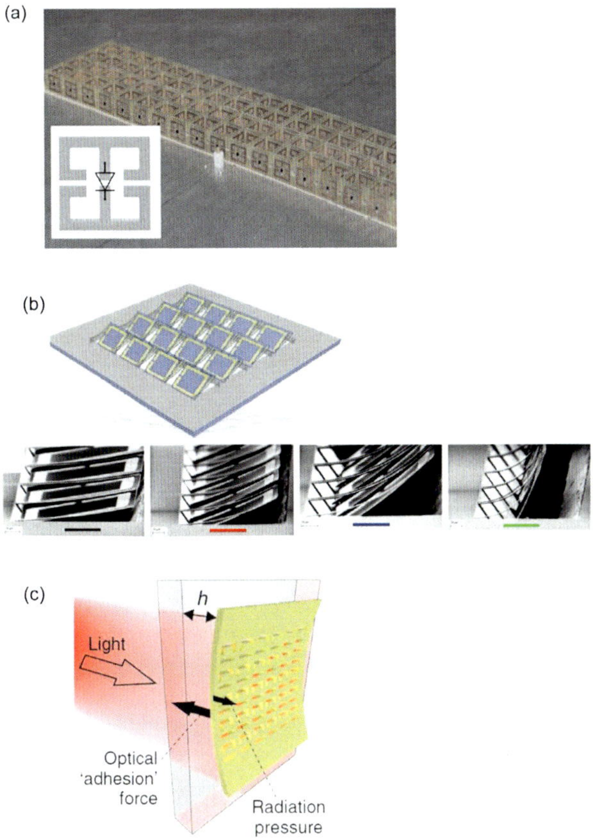

Figure 1.21 Metadevices collections. (a) Nonlinear electric metadevice composed of varactor diodes at GHz. The inset illustrates the unit cell. (b) Reconfigurable metadevice at THz, controlled by RTA. (c) Optical-force-driven metadevice at a NIR range. (a) was reprinted with permission from Powell et al. (2009). Copyright 2009, AIP Publishing LLC. (b) and (c) were reprinted with permission from Tao et al. (2009) and Zhang et al. (2012), respectively; copyright 2009 and 2012 by the American Physical Society, respectively.

The tilting angle can be controlled in a reconfigurable manner. At the lower panels, SEM images of as-fabricated samples are shown at the most left-hand side with black scale bar of 20 μm; rapid thermal annealing (RTA) at 350, 400, and 450°C resulted in increase in tilting angles from the second-left to most right panels with red, blue, and green scale bars of 20, 15, and 20 μm, respectively. By changing the tilting angles from the substrate, the SRRs show different resonant response to the illuminating THz waves traveling normally to the substrate because incident angles to the SRRs were changed. Of course, combination of MMs with MEMS techniques make reconfigurable MMs and indeed has been conducted in several cases to date.

Figure 1.21c shows a conceptual illustration of optical-force-driven metadevices working at optical frequencies in a NIR range (Zhang et al., 2012). It was numerically shown that a MM of perforated Au film with rectangular holes exhibits adhesive force under light illumination on the MM resonance. The illumination light gives radiation pressure to the MM. However, it was turned out that the adhesive force can be several times larger than the radiative force. This situation was conceived from the van der Waals force on the gecko toes. Readers who are interested in metadevices can see many examples in a review (Zheludev and Kivshar, 2012).

Metasurfaces Metasurfaces were stimulated by MMs and recently became popular by a study on single-layered periodic structure of a set of Au V-shape-based antenna array (Yu et al., 2011). Figure 1.22 shows a schematic illustration of the metasurface. The single-layer metasurface induced *anomalous* refraction (dashed arrow) to incidence (arrow), and greatly stimulated the studies on single-layered artificial micro- or nanostructures. One of the features in the single-layer metasurface is that refraction is cross polarized to the incident polarization; that is, when incidence is s-polarized, the refraction is p-polarized, and vice versa. The polarization responses were explained from the phase modulation of the refracted beam. For the working wavelengths of 8 μm, the periodicity (both-ends arrow) was varied from 11 to 21 μm (Yu et al., 2011). Note that the in-plane structure was not a subwavelength structure but a diffractive element. Thus, metasurfaces are not limited to subwavelength structures; this point is different from MMs that usually assume subwavelength periodicity to the working wavelengths. Taking account of these points, it seems possible to interpret the metasurface in Fig. 1.22 as a diffraction-controlling structure.

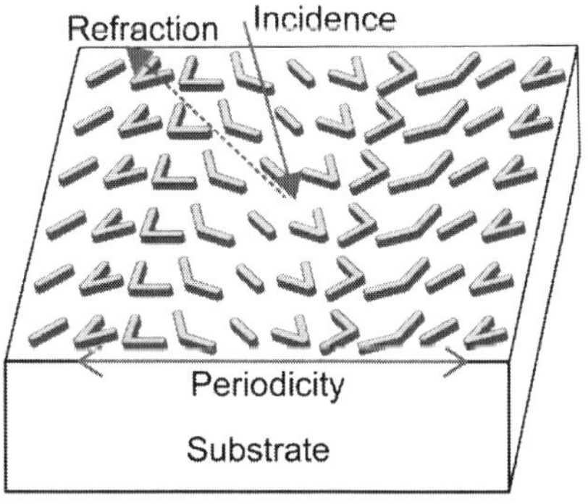

Figure 1.22 Schematic drawing of a single-layer metasurface of metallic V-shape-based antenna array, based on (Yu et al., 2011).

One of the key concepts in metasurfaces is to realize light manipulation using *flat* optical structures. Along the concept, several applications such as holography (Zheng et al., 2015), light focusing (Aieta et al., 2012), and so on are being explored.

In addition to these purely optical applications, enhanced molecular spectroscopy handling light–matter interaction in a desired way and IR emitters converting electric power to IR light have been demonstrated as described in Chapter 5.

Extended notion of MMs As is well known, thermodynamics, acoustics, and elastics have established fundamental differential equations, which include material parameters. In this feature, these disciplines are in common with electromagnetism. MMs addressed so far were originally intended to produce arbitrary materials parameters on electromagnetism in artificial ways. Similar scenario holds to thermodynamics, acoustics, and elastics; indeed, many reports on cloaking of thermal diffusion and of acoustic and elastic waves have already supported the scenario (Kadic et al., 2013). However, it would be fair to point out that most of the cloaking demonstrations were limited to 2D cases and that it is not easy to realize perfect three-dimensional (3D) cloaking for the classical waves though they allows much larger unit cell in size than those in photonic MMs.

1.5 Numerical Methods

1.5.1 *Rigorous Coupled-Wave Analysis (RCWA) and Scattering (S) Matrix*

Most of precisely fabricated nanostructures are composed of periodic structures. Periodic array of plasmonic resonators are also produced as plasmonic crystals (PlCs) and MMs. In these periodic structures, Fourier transformation is a valid computational method to examine the EM eigenmodes accurately. Generally, Fourier transformation is applied to periodic systems based on the mathematical ground. One of the advantages in the Fourier transformation is that diffractive components R_{mn} and T_{mn} by the periodic structures are directly evaluated, where R_{mn} and T_{mn} denote the *mn*-th reflective and transmissive components in diffraction, respectively. This feature is distinct from other methods described later (Sections 1.5.2 and 1.5.3). Here we survey how to implement the Fourier modal method or the RCWA method in electromagnetism. As terminology, RCWA method is now more popular term than Fourier modal method; accordingly, we call the computational method RCWA method.

Inverse rule Let us first consider a general case that a product of complex-valued 1D periodic functions, $f(x)$ and $g(x)$, with periodicity d is equal to $h(x)$, that is, $f(x+d) = f(x), g(x+d) = g(x)$, and $h(x) = f(x)g(x)$. The Fourier coefficients satisfy with

$$h_n = \sum_{m=-\infty}^{+\infty} f_{n-m} g_m. \tag{1.67}$$

Note that the functions f and g are assumed to be finite-valued (or $|f|, |g| < \infty$) without loss of generality for the present purpose. Practically, numerical calculations require to truncate the sum in Eq. 1.67 at a finite order such that

$$h_n^{(M)} = \sum_{m=-M}^{+M} f_{n-m} g_m. \tag{1.68}$$

The issue in Fourier transformation in practical computations is whether the following relation holds or not:

$$h_n^{(M)} \approx h_n \tag{1.69}$$

at a certain M that is not extremely large.

Unfortunately, Eq. 1.69 does not hold for periodic nanostructures including metals, which prevented the numerical studies on periodic metallic nanostructures including deep gratings for about thirty years (concretely, from about 1970 to 1996). The bad convergence had been a puzzle in the electromagnetism computation. The first key solution was independently invented by the three groups at 1996 (Lalanne and Morris, 1996; Li, 1996b; Granet and Guizal, 1996). The factor f_{n-m} in Eq. 1.68 is also represented in a matrix form, which is defined by $[[f]]_{n,m} = f_{n-m}$. With this notation, the solution for the puzzle was given by the formula:

$$\hat{h}_n^{(M)} = \sum_{m=-M}^{+M} [[1/f]]_{n,m}^{-1} g_m, \tag{1.70}$$

instead of Eq. 1.68. Equation 1.70 is called inverse rule in Fourier factorization and satisfies with $\hat{h}_n^{(M)} \to h_n$ $(M \to \infty)$ uniformly at almost everywhere in mathematics, which supports practical relation such that

$$\hat{h}_n^{(M)} \approx h_n \tag{1.71}$$

at an executable M.

Formulation by Li Next, we briefly describe how to apply the inverse rule to actual problems including plasmonic resonators. The general formulation for 2D systems was shown by Li (Li, 1997). Optical configuration is shown in Fig. 1.23a. In the periodic layer of finite thickness (gray), permittivity is periodic in the xy plane such as $\varepsilon(x + d_x, y + d_y, z) = \varepsilon(x, y, z)$ for periodicity d_x and d_y.

Assuming the monochromatic temporal dependence such that

$$\mathbf{E}(\mathbf{r}, t) = \mathbf{E}(\mathbf{r}) \exp(-i\omega t), \tag{1.72}$$

$$\mathbf{H}(\mathbf{r}, t) = \mathbf{H}(\mathbf{r}) \exp(-i\omega t), \tag{1.73}$$

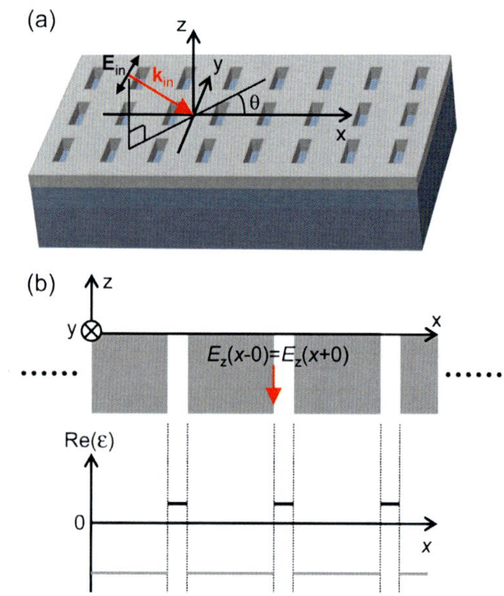

Figure 1.23 (a) Schematic of a periodic structure in the xy plane, composed of a metallic film of rectangular air-hole array (gray). The periodic structure has finite thickness along the z axis and is set on a substrate (pale blue). (b) Upper: The xz section of the periodic layer is shown. Lower: The discontinuous permittivity distribution at a frequency is schematically shown along the x axis.

Maxwell equations in Eqs. 1.7 and 1.8 are explicitly written for the periodic layer as follows.

$$\partial_y E_z - \partial_z E_y = i\omega\mu_0 H_x \tag{1.74}$$

$$\partial_z E_x - \partial_x E_z = i\omega\mu_0 H_y \tag{1.75}$$

$$\partial_x E_y - \partial_y E_x = i\omega\mu_0 H_z \tag{1.76}$$

$$\partial_y H_z - \partial_z H_y = i\omega\varepsilon_0\varepsilon(x, y, z) E_x \tag{1.77}$$

$$\partial_z H_x - \partial_x H_z = i\omega\varepsilon_0\varepsilon(x, y, z) E_y \tag{1.78}$$

$$\partial_x H_y - \partial_y H_x = i\omega\varepsilon_0\varepsilon(x, y, z) E_z \tag{1.79}$$

Note that we omit $(\mathbf{r})$ for the EM field components in Eqs. 1.74–1.79. In the periodic layer, the following Floquet-Fourier expansion is valid:

$$E_x(\mathbf{r}) = \sum_{m,n} E_{xmn}(z)\exp(i\alpha_m x + i\beta_n y), \tag{1.80}$$

$$\alpha_m = \alpha_0 + \frac{2\pi m}{d_x}, \tag{1.81}$$

$$\beta_n = \beta_0 + \frac{2\pi n}{d_y}, \tag{1.82}$$

where m and n are integers, and α_0 and β_0 are the x and y components of incident wavevector, respectively. Similar expansions hold for other field components (E_y, H_z, etc.). The relative permittivity in the periodic layer is also expanded such that

$$\varepsilon(x, y, z) = \sum_{m,n} \varepsilon_{mn} \exp[2\pi i(m/d_x + n/d_y)]. \tag{1.83}$$

Note that ε_{mn} is defined in each layer; in this sense, ε_{mn} depends on z.

To obtain eigen equations for the periodic layer using only E_x, E_y, H_x, and H_y, let us eliminate E_z component. Equation 1.79 is modified as

$$E_z = -\frac{1}{i\omega\varepsilon_0\varepsilon(x, y, z)}(\partial_x H_y - \partial_y H_x). \tag{1.84}$$

Note that the relative permittivity ε always satisfies $\varepsilon \neq 0$. Equation 1.84 exhibits a continuous function at the left-hand side as shown in Fig. 1.23b and a product of discontinuous functions at the right-hand side. Thus, the inverse rule in Eq. 1.70 has to be applied in taking the Fourier-coefficient factorization.

By substituting Eq. 1.84 into Eq. 1.75, we obtain

$$\partial_z E_x = i\omega\mu_0 H_y - \frac{1}{i\omega\varepsilon_0}\partial_x \left[\frac{1}{\varepsilon(x, y, z)}(\partial_x H_y - \partial_y H_x)\right]. \tag{1.85}$$

This equation is represented with Fourier coefficients by using the inverse rule:

$$\frac{\omega\varepsilon_0}{i}\partial_z E_{xmn} = k_0^2 H_{ymn} - \alpha_m \sum_{j,l} [[\varepsilon]]^{-1}_{mn,\,jl}(\alpha_j H_{yjl} - \beta_l H_{xjl}) \tag{1.86}$$

where k_0 denotes wavenumber in vacuum, defined by $k_0 = \omega/c_0$. Note that we have extended the matrix notation $[[\cdot]]$ for 2D case such that

$$[[\varepsilon]]_{mn,\,jl} = \varepsilon_{m-n,\,j-l}.$$

Similar derivation leads to other equations with Fourier representation as follows.

$$\frac{\omega\varepsilon_0}{i}\partial_z E_{ymn} = -k_0^2 H_{xmn} \\ -\beta_n \sum_{j,l} [[\varepsilon]]^{-1}_{mn,jl}(\alpha_j H_{yjl} - \beta_l H_{xjl}) \quad (1.87)$$

$$\frac{\omega\mu_0}{i}\partial_z H_{xmn} = \alpha_m(\alpha_m E_{ymn} - \beta_n E_{xmn}) \\ -k_0^2 \sum_{j,l} \lceil\lfloor\varepsilon\rfloor\rceil_{mn,jl} E_{yjl} \quad (1.88)$$

$$\frac{\omega\mu_0}{i}\partial_z H_{ymn} = \beta_n(\alpha_m E_{ymn} - \beta_n E_{xmn}) \\ +k_0^2 \sum_{j,l} \lfloor\lceil\varepsilon\rceil\rfloor_{mn,jl} E_{xjl} \quad (1.89)$$

The symbols $\lceil\lfloor\cdot\rfloor\rceil$ and $\lfloor\lceil\cdot\rceil\rfloor$ were defined in the following. In the periodic layers, relative permittivity can be expanded using coefficient matrices $\lceil\cdot\rceil$ and $\lfloor\cdot\rfloor$ such that

$$\varepsilon(x, y, z) = \sum_{m,j} \lceil\varepsilon\rceil_{mj}(y, z)\exp[2\pi i(m - j)x/d_x], \quad (1.90)$$

$$\varepsilon(x, y, z) = \sum_{n,l} \lfloor\varepsilon\rfloor_{nl}(x, z)\exp[2\pi i(n - l)y/d_y]. \quad (1.91)$$

Equations 1.90 and 1.91 are definitions of $\lceil\cdot\rceil$ and $\lfloor\cdot\rfloor$. The symbols $\lceil\lfloor\cdot\rfloor\rceil$ and $\lfloor\lceil\cdot\rceil\rfloor$ were given by

$$\lceil\lfloor\varepsilon\rfloor\rceil_{mn,jl} = \lceil(\lfloor 1/\varepsilon\rfloor^{-1})_{nl}\rceil_{mj}, \quad (1.92)$$

$$\lfloor\lceil\varepsilon\rceil\rfloor_{mn,jl} = \lfloor(\lceil 1/\varepsilon\rceil^{-1})_{mj}\rfloor_{nl}. \quad (1.93)$$

The inverse rule in the Fourier factorization enabled practical truncation in Eq. 1.70, whereas the Maxwell equations became quite complicated form as shown in Eqs. 1.86–1.89. These equations are summing up in a matrix form:

$$\partial_z \begin{pmatrix} \langle E_x\rangle \\ \langle E_y\rangle \\ \langle H_x\rangle \\ \langle H_x\rangle \end{pmatrix} = \begin{pmatrix} 0 & F \\ G & 0 \end{pmatrix} \begin{pmatrix} \langle E_x\rangle \\ \langle E_y\rangle \\ \langle H_x\rangle \\ \langle H_x\rangle \end{pmatrix} \quad (1.94)$$

where the symbol $\langle\cdot\rangle$ denotes the vector $(\cdot)_{mn}$ arrayed in Toeplitz order, and F and G are submatrices composed of $[[\varepsilon]]_{mn,jl}$, α_m, β_n, etc. Assuming that the z-dependence of the vector components is exponential, for example, $E_{xmn}(z)$ is

$$E_{xmn}(z) = E_{xmn} \exp(i\gamma_{mn}z), \tag{1.95}$$

Eq. 1.94 becomes an eigenvalue equation with replacing ∂_z with $i[\gamma]_{mn}$ where $[\cdot]$ denotes diagonal matrix in the Toeplitz order. To save computational resource, Eq. 1.94 is reduced such that

$$\left(FG + [\gamma^2]\right)\begin{pmatrix}\langle E_x\rangle \\ \langle E_y\rangle\end{pmatrix} = \vec{0}. \tag{1.96}$$

Note that the matrix in Eq. 1.94 has the size of $4(2M+1)^2 \times 4\,(2M+1)^2$ while the matrix in Eq. 1.96 has that of $2(2M+1)^2 \times 2\,(2M+1)^2$, reduced by 4 times, where M denotes truncation order in the Fourier expansion.

Overall, the solution of the eigenvalue equation in Eq. 1.96 determines EM eigenmodes in periodic layers, expressed by Eq. 1.80. At the end, we mention that the RCWA is now a popular method in computing optical spectra of periodic structures and available as commercial packages.[f]

S-matrix algorithm To implement calculations for optical spectra and EM fields of plasmonic resonators, a more elaborate algorithm is needed, which is S-matrix algorithm. We briefly mentions it before showing an actual computational result.

Before describing the S-matrix algorithm, we mention transfer matrix (T matrix or T_{mat}) that is more easily understood by intuition. Figure 1.24 shows a simple configuration including single-layered periodic structure. The T_{mat} in Fig. 1.24 is explicitly expressed as

$$T_{\text{mat}} = I_{21}\begin{pmatrix}[\exp(-i\kappa d)] & 0 \\ 0 & [\exp(i\kappa d)]\end{pmatrix} I_{10} \tag{1.97}$$

where the symbol $[\cdot]$ denotes diagonal matrix and κ represents boost along the z axis, satisfying with $\text{Im}(\kappa) \geq 0$. The matrices I_{kj} are interface matrices at the interface of the layers k and j. In the T_{mat}, the input components are incidence $\langle \text{in}\rangle$ and reflection $\langle R\rangle$ while the output are transmission $\langle T\rangle$ and zero vector $\langle 0\rangle$. They are connected such that

[f]For example, DiffractMOD, http://www.optics.synopsys.com

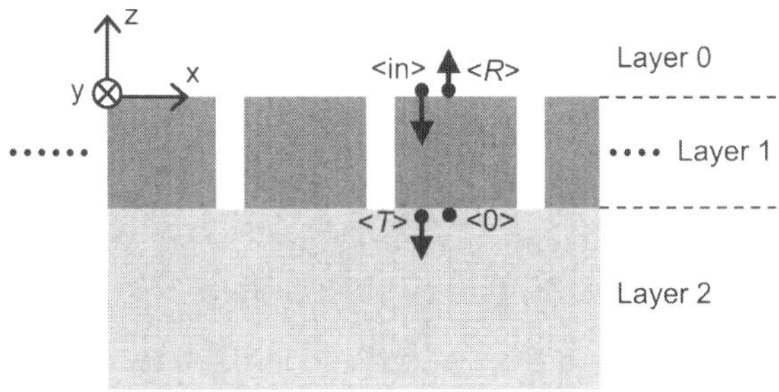

Figure 1.24 Scattering configuration. Incidence is symbolized with $\langle in \rangle$. Reflection and transmission components are represented with $\langle R \rangle$ and $\langle T \rangle$, respectively. Symbol $\langle 0 \rangle$ denotes a null component.

$$\begin{pmatrix} \langle T \rangle \\ \langle 0 \rangle \end{pmatrix} = T_{\text{mat}} \begin{pmatrix} \langle \text{in} \rangle \\ \langle R \rangle \end{pmatrix}. \tag{1.98}$$

General periodic structures are not single-layered but are composed of stacked layers as shown in Fig. 1.20. Eigenmodes in each layer expressed by Fourier-coefficient vectors are connected at the interfaces by matrix multiplication. By combining the interface connection and boost in each layer, one naturally reaches T-matrix formula (Markoš and Soukoulis, 2008). In the layer including metals of complex permittivity, the κ is complex-valued, so that either $\exp(i\kappa d)$ or $\exp(-i\kappa d)$ is decreasing and the other is exponentially growing. The latter easily results in numerical divergence in computations; practically, the T_{mat} formula is not useful, except for the cases that κ is purely real-valued.

To avoid the numerical divergence, the S-matrix method is employed. S matrix itself is a standard theoretical tool in physics. Input and output components are connected by S matrix in any physical phenomena. Figure 1.25 shows how the S matrix (S) relates the input ($\langle \text{in} \rangle$ and $\langle 0 \rangle$) to output ($\langle R \rangle$ and $\langle T \rangle$). In the matrix form

$$\begin{pmatrix} \langle T \rangle \\ \langle R \rangle \end{pmatrix} = S \begin{pmatrix} \langle \text{in} \rangle \\ \langle 0 \rangle \end{pmatrix}. \tag{1.99}$$

Obviously, the input and output components were just reconfigured; therefore, Eq. 1.99 is mathematically equivalent to the T-matrix formulation.

In the RCWA method, the symbols $\langle \cdot \rangle$ represent vectors composed of the Fourier coefficient on EM field. The boost matrix is

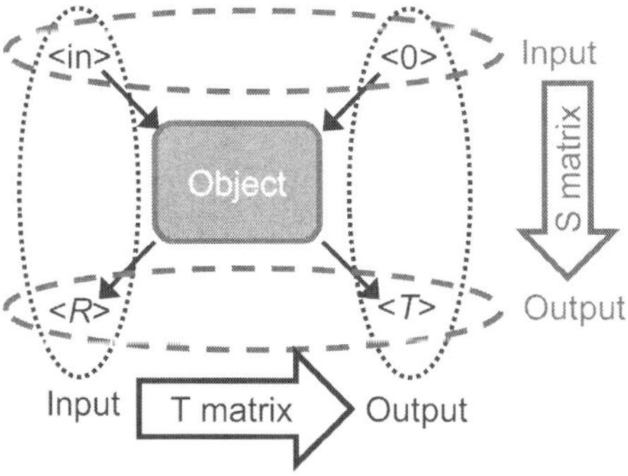

Figure 1.25 Reconfiguration from T-matrix to S-matrix representations. The set of input and output components in T-matrix algorithm, indicated by dotted circles, is changed into the set in the S-matrix algorithm, indicated by dashed circles.

reformed such as

$$\begin{pmatrix} [\exp\{-i\kappa(-d)\}] & 0 \\ 0 & [\exp(i\kappa d)] \end{pmatrix}. \tag{1.100}$$

Note that the boost matrix in Eq. 1.100 does not at all include the exponentially growing components $\exp(-i\kappa d)$ where $\mathrm{Im}(\kappa) \geq 0$. Therefore, the S-matrix method is numerically stable.

In general, the S-matrix algorithm was derived from T-matrix algorithm by recurrent formula (Ko and Inkson, 1988; Li, 1996a). The derivations were independently conceived for different aims: the former was to solve electronic transport in quantum wells of semiconductors as an issue in quantum mechanics (Ko and Inkson, 1988) and the latter was to calculate light propagation in periodic media as issue in classical electromagnetism (Li, 1996a). The S-matrix algorithm for electromagnetism was explicitly derived in 1996 (Li, 1996a). By combining the RCWA method and the S-matrix algorithm, arbitrary stacking 2D periodic structures, which practically mean 3D periodic structures in experiment, are analyzed in a first-principles way in classical electromagnetism. The inverse rule in Fourier factorization makes the computation practical.

Numerical calculations in this book We show many optical spectra such as R and T calculated by the RCWA code prepared by ourselves; the code incorporated the S-matrix algorithm (Li, 1996a). We refer to the code as RCWA+S-matrix code from now on. The code was optimized for the vector-oriented supercomputers in a multi-parallel implementation manner. Also, EM field distributions are output based on the RCWA+S-matrix method.

To ensure realistic computations, we use materials parameters from literature compiling measured values (Rakić et al., 1998; Palik, 1991) or widely known representative values.

Precision in the RCWA+S-matrix method For 2D periodic structures shown from now on, the truncation order M was set to be 20 or more for each x an y component; the indices m and n run such as

$$m, n = -M, -M+1, \cdots, -1, 0, 1, \cdots, M-1, M$$

in Eq. 1.80. Then, estimated numerical fluctuations of the output quantities (R_{mn} and T_{mn}) were about within 1%. For 1D periodic structures, one can assume that d_y is infinity in Eq. 1.82; as a result, requirements in numerical implementation become much less than 2D cases. It is therefore possible to set large truncation order such as $M = 100$ and to suppress numerical fluctuations less than 0.5%.

The numerical precision is connected to the memory in the numerical implementations. Although it was sometimes said that the RCWA method requires large computation resources in comparison with the FDTD method (see Section 1.5.2), the amount of resources, to our knowledge, is essentially dependent on how to save the memory in the codes. In fact, we do not find any disadvantage in the RCWA method in terms of the required resources.

We also mention the isolated systems that are not periodic systems. The RCWA+S-matrix method was originally formulated for periodic systems (Li, 1997, 1996a) and therefore is less suitable for the isolated systems. Still, it is able to calculate for the isolated systems in an approximate way. Specifically, by setting the periodicity to be long enough to neglect interplay and interference between the unit element, the EM field distributions evaluated by the RCWA+S-matrix becomes equivalent to the isolated systems. In other numerical techniques such as FDTD, perfect matching boundary condition is assigned to absorb the scattered field; thus, the computation procedures are also approximate ways.

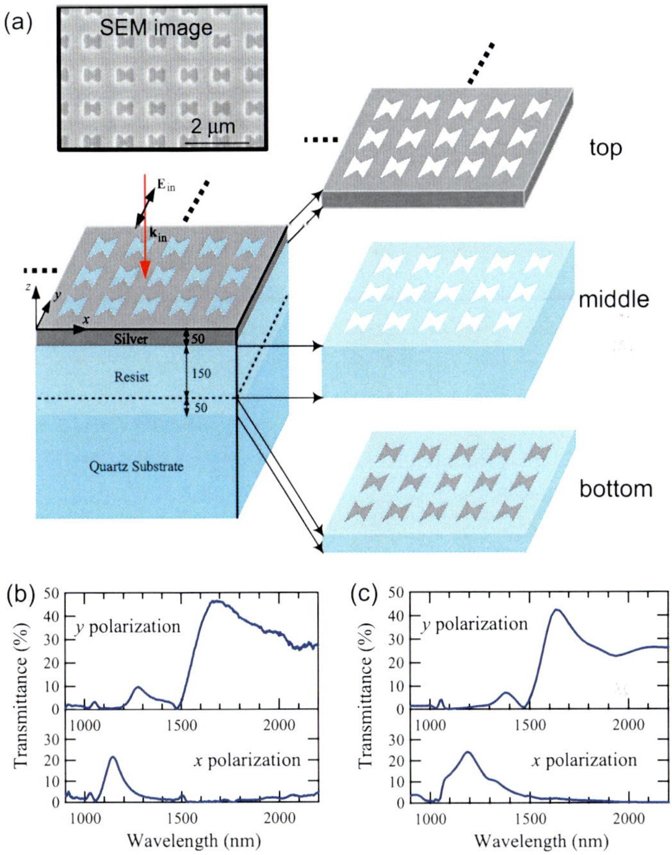

Figure 1.26 Butterfly-type SC PlC. (a) SC structure composed of top, middle, and bottom layers. The length is shown in the units of nm. The inset is the top-view SEM image. Scale bar (black) indicates 2 μm. (b) and (c) Measured and calculated T spectra, respectively. T spectra at *y* polarization are displayed in the upper and T at *x* polarization in the lower. The calculation was implemented by using the RCWA+S-matrix method. Adapted with permission from Iwanaga (2010a). Copyright 2010, AIP Publishing LLC.

An example of the RCWA+S-matrix method Figure 1.26 shows an actual example of the RCWA+S-matrix. Figure 1.26a displays schematic of a 2D periodic structure composed of three stacked layers (Iwanaga, 2010a). The top layer comprised perforated metallic film. The shape of air hole is butterfly-shaped. The metal was Ag. The bottom layer also included Ag tips, which were same shape to the air hole in the top layer. The middle layer consisted of perforated dielectric film. The transparent dielectric was resist of relative permittivity 2.4025 in the wavelength range of interest, used in EBL. The thickness was 50, 150, and 50 nm from the top to bottom layers.

In terms of metallic nanostructures, the top and bottom layers are complementary. We thus name the stacked structure stacked complementary (SC) structure. The SC structure was made on quartz substrate of relative permittivity 2.1316. The inset shows the top-view SEM image of the fabricated sample. The scale bar (black) indicates 2 μm.

Figures 1.26b and 1.26c show measured and calculated T spectra, respectively. The calculation was implemented with the RCWA+S-matrix. The T spectra exhibit distinct shapes at x and y polarizations. In particular, large T was observed at 1600 nm with y polarization while T at 1600 nm with x polarization was very small; the extinction ratio T_y/T_x exceeded 1500 in experiment. Thus, it was confirmed in the SC PlC that measured and calculated T spectra are in good agreement with each other. The plasmonic resonances associated with the T peaks have been numerically clarified from the EM-field distributions (Iwanaga, 2010b,c).

1.5.2 *Finite-Difference Time-Domain (FDTD) Method*

FDTD method is an established technique for EM-wave analysis, which has been applied to analyses for antenna, electric circuits, and so on. The first application to actual problem was reported in 1960s (Yee, 1966). In the FDTD method, Maxwell equations are approximated in the time domain; for example, the z component of Eq. 1.7 is expressed such that

$$\frac{E_x(x, y+\Delta_y/2, z, t) - E_x(x, y-\Delta_y/2, z, t)}{\Delta_y} - \frac{E_y(x+\Delta_x/2, y, z, t) - E_y(x-\Delta_x/2, y, z, t)}{\Delta_x} = -\frac{H_z(x, y, z, t+\Delta_t/2) - H_z(x, y, z, t-\Delta_t/2)}{\Delta_t} \quad (1.101)$$

where Δ_i denotes finite difference and μ is set to unity for simplicity.

The FDTD codes at optical frequencies were extensively developed for photonic crystals that consist of loss-free dielectric or semiconductors (Johnson and Jannopoulos, 2001); therefore, it was assumed that the employed permittivity is purely real. As is clear from the basic equation such as Eq. 1.101, the FDTD method is suitable to evaluate transient motion of EM waves and to output visually appealing animations of EM waves.

The FDTD method is also a well-used method for plasmonic resonators. In particular, The method was extensively applied to metallic nanoparticles, the assembles, and so on around 2000–2010 (for example, Lakowicz et al., 2008). This was probably due to the following reasons.

(i) At the early stage of plasmonics, major research targets were isolated metallic nanoparticles and the assembles (Figs. 3.5–3.13). Therefore, the resonant near-fields of the structures attracted the greatest interest. To meet the interest, the FDTD method was suited.
(ii) For the tiny structures, electric dipoles were often convenient as an EM-wave source. The point source can be naturally incorporated in the FDTD method.[g]
(iii) From practical point of view, the FDTD method was developed since 1960s and available at the time. In addition, there was no other computational method that was widely available as commercial softwares.[h] The RCWA method in Section 1.5.1 gradually became popular around 2010.

In comparison with the RCWA method that was originally intended to calculate for metallic nanostructures, the FDTD method is originally less compatible and requires additional algorithm to incorporate complex permittivity, so that the simplifications, which for example assumed the metal to be a perfect electric conductor, were often introduced as an additional assumption (Rodrigo, 2012).

From the results on plasmonic resonators at IR ranges (for example, Ameling and Giessen, 2010), it now seems possible

[g]The incident plane waves in the FDTD method are prepared using the dipoles aligned on a line.

[h]For example, CST studio suite, http://www.cst.com

to incorporate measured permittivity of constituent materials in commercial softwares. In evaluating the optical spectra such as R and T spectra, the FDTD method usually leaves detectable fringes in the frequency domain, due to the Fourier transformation from the time domain (Rodrigo, 2012; Ameling and Giessen, 2010), whereas the RCWA method is in principle free from such artifacts.

1.5.3 *Finite-Element Method (FEM)*

FEM has been developed to resolve partial differential equations appearing mathematics and physics. For electromagnetism, the formulation was reported was developed in 1990s (as a review, Volakis et al., 1994). One of the features in the FEM is to be able to divide constituents and media by grids of arbitrary dimensions.

Figure 1.27 shows an example of spatial grid generated by a FEM software. Figure 1.27 (left) is a 3D view of the calculated domain, which was unit in calculating for the SC structure like Fig. 1.26. Periodic condition was assigned to the *xz* and *yz* boundaries in Fig. 1.27. For clarity, the grids only on the interfaces are shown in Fig. 1.27; the grids were triangle-shaped and produced semi-automatically in

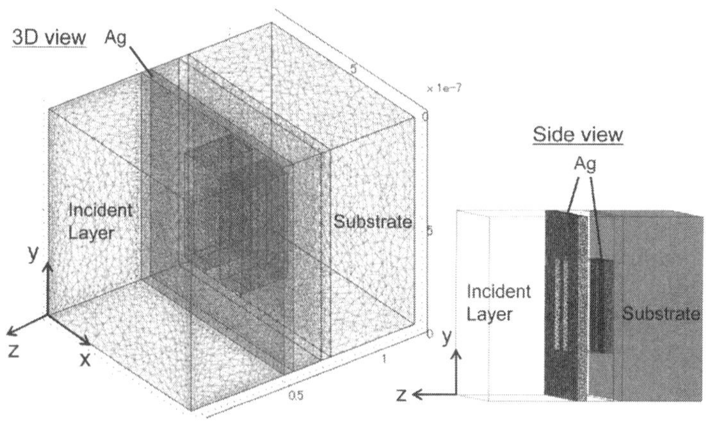

Figure 1.27 Spatial grid semi-automatically generated for the FEM computation using a commercial software (see the footnote in this subsection). 3D oblique view (left) and side view (right) are shown. In the latter, the metallic domains of Ag are only divided by grids for clarity.

accordance with the user-defined size by a commercial software (COMSOL Multiphysics). The incident layer, the SC structure, and the substrate are located from left to right. The incident wave was set to come through the *xy* plane at the left-hand side. Figure 1.27 (right) shows a side view of spatial grids only in metallic nanostructures. Obviously, the grids were densely produced; they are triangles on the surface and tetrahedrons in the nanostructures. The FEM computations were implemented to resolve the resonant modes in the SC PlC and revealed the eigenmodes successfully (Iwanaga, 2010b,c); the remarkably anisotropic plasmonic resonances are described in Section 3.8.

To keep precision in the implementations at a good level, transparent media were divided into the dimensions less than 1/30 effective wavelength, which is defined by λ/n (λ: wavelength in vacuum, n: refractive index of the transparent media). As for metals, much finer grids are needed. Skin depth of metals at optical wavelengths is a few tens of nm; therefore, grids of sides of a few nm or less were set in Fig. 1.27. Such fine grids requires large memory in the implementations and easily exceeded 100 GB. As for 1D structures, the unit domain is 2D and requires much less memory. Accordingly, computation time is much shorter. Nowadays, FEM-coded commercial softwares are available for computing EM-field distributions.[i] In this point, the FEM plays a similar role to the FDTD method (Section 1.5.2). On the other hand, FEM is less suitable than the RCWA+S-matrix (Section 1.5.1) in computing optical spectra in the frequency domain because diffractive components generally have to be extracted using additional algorithm that is not explicitly implemented in the FEM.

1.6 Nanofabrication Methods

There are several techniques to fabricate nanostructures by top-down approach. Old one is photolithography incorporating interference effect of light waves. Around 2000, FIB technique and EBL gradually became standard in top-down nanofabrication. The

[i] For example, COMSOL Multiphysics, http://www.comsol.com

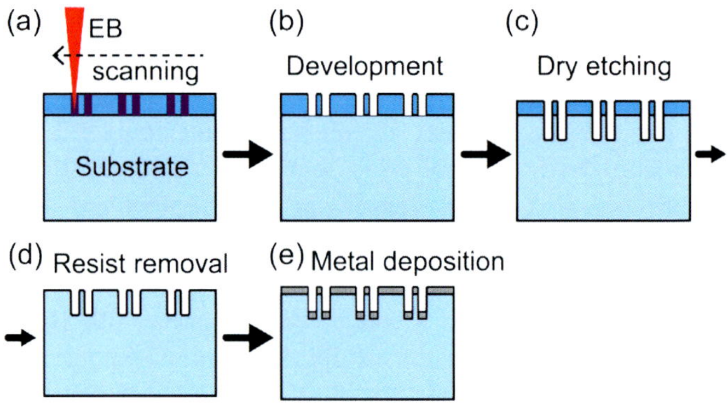

Figure 1.28 A standard EBL procedure. (a) EB writing in posi-resist (sky blue). (b) Development of the EB pattern. (c) Dry etching of substrate (pale blue). (d) Resist removal. (e) Metal (gray) deposition. The procedure results in a SC PlC described in Section 3.8.

former was a direct fabrication technique by use of ion beam and has been employed to dig holes and to cut specimen in micro- and nano-dimensions; the latter is summarized as follows.

EBL procedure EBL was developed in accordance with the needs in semiconductor industry. The nanostructures were initially made of semiconductors such as Si and were called photonic crystals (Johnson and Jannopoulos, 2001; Lourtioz et al., 2005). Metallic nanostructures were also made in a similar procedure to that for semiconductors and opened a new trend, called plasmonics (Maier et al., 2001; Maier and Atwater, 2005; Murray and Barnes, 2007).

Figure 1.28 shows a standard EBL procedure. Resist for EB is spin-coated and usually prebaked in advance; (a) EB-writing in accordance with designs is executed; (b) development is done and the EB-writing patterns are removed for posi-type resists; (c) dry etching using plasma gas is done for the materials under the resist; (d) the EB resist is removed; (e) metal is deposited perpendicularly to the surface. Conducting the sequence above, we obtained a SC PlC of unit cell including II shape, which will be described in Section 3.8.

Periodic patterns of submicron periodicity can also be fabricated by interference photolithography though the patterns are limited to simple square or hexagonal lattices (for example, Okamoto et al.,

2004; Zhang et al., 2005). When one seeks more large degree of freedom in nanofabrications, EBL will be the first choice. Besides, we mention a recent trend; recently, extreme UV (EUV) lithography has developed and is currently used in the semiconductor industry for producing computer tips.

Nanoimprint lithography (NIL) procedure NIL has been developed in this decade. The first report appeared in 1995 (Chou et al., 1995), demonstrating 25-nm-diameter perforation by direct imprinting technique. The direct imprinting technique is valid even for sub-10 nm scale (Chou et al., 1997). By combining the imprinting with various lithographic techniques, high-throughput, large-area lithography has been available to date. At present, there are two NIL techniques; one is thermal NIL employed in the first demonstration above and the other is UV NIL using resists cured under UV irradiation.

Here we describe a procedure in UV NIL, which is illustrated in Fig. 1.29:

(a) a substrate is coated with resist and prebake is done;
(b) the substrate is imprinted by a quartz mold and irradiated by UV light;
(c) after removal of the mold, thin residual resist remains at the bottom of the resist layer;
(d) after conducting postbake, the residual thin film was etched and removed with a slow-rate and tunable treatment using $O_2 + N_2$ gas plasma;
(e) the substrate is etched down in an anisotropic manner to a designed depth;
(f) the resist is removed by a wet or dry process;
(g) metal is almost normally deposited onto the nanopatterns.

The SC PlCs produced through the procedure are described later (Section 3.8).

Figure 1.30a shows a typical photograph of a nanoimprinted substrate of 2 inch diameter. Four areas, which look blue due to diffraction, comprise periodic nanostructures and are 1×1 cm^2 dimension.

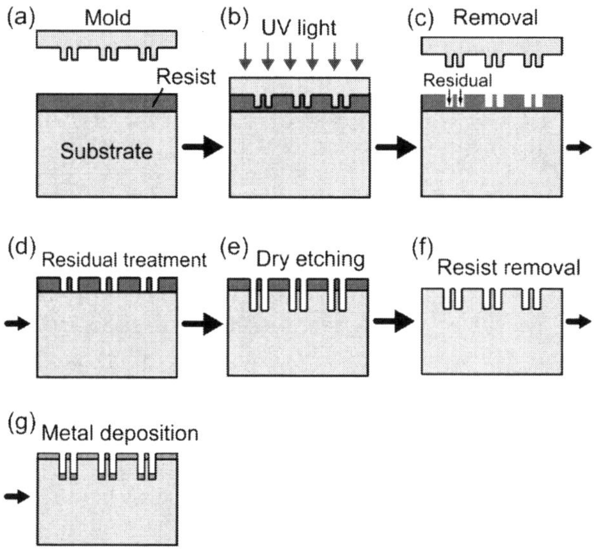

Figure 1.29 UV NIL procedure for a SC PIC. (a) Resist for UV NIL was coated on substrate. (b) Press of mold on the resist-coated substrate and UV-light irradiation. (c) Removal of resist. Thin residual film is left as indicated by arrows. (d) Residual film is removed by slow-rate etching with $O_2 + N_2$ gas plasma. (e) Dry etching for the substrate using the resist mask. (f) Resist removal. (g) Metal is normally deposited and a SC PIC is obtained.

Figure 1.30b displays a section-view SEM image of a nanoimprinted resist pattern of II-shape array and just corresponds to the step (c) in Fig. 1.29. White scale bar indicates 1 μm. Obviously, the nanoimprinted pattern is highly uniform in structure.

Thermal NIL is another major NIL technique which uses heating procedure instead of the UV exposure in Fig. 1.29b. Other steps in the thermal NIL procedure are similar to those in UV NIL procedure. A high-precision thermal NIL was reported on patterning of submicron Si structures on the whole 4-inch substrate by conducting two-times nanoimprinting (Wang et al., 2014). Currently, NIL techniques are developing to a few directions

(i) One of them is to produce nanostructures at large-area as possible, using role-to-role imprinting with a high speed up to 20 m/min; at present, microstructures of periodicity more

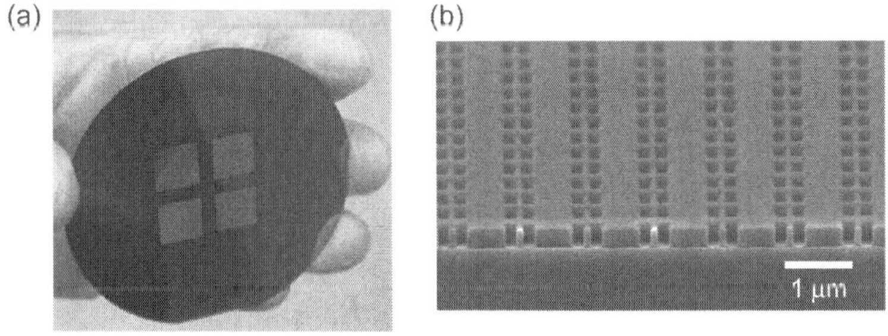

Figure 1.30 (a) Photograph of a nanoimprinted substrate. Four squares of $1 \times 1\ cm^2$ are located on a 2 inch wafer. Owing to the courtesy by Dr. B. Choi. (b) Oblique-view SEM image of an UV nanoimprinted resist pattern of II-shape array. The image corresponds to the step (c) in Fig. 1.29. White scale bar indicates 1 μm.

than 50 μm are mainly produced (Mäkelä et al., 2011; Mäkelä and Haatainen, 2012). This technology will enable to send products made of microstructure-patterned films to the market for display in the near future.

(ii) One of others is to produce many nano- and microstructured elements in a parallel way through one NIL process (Ofir et al., 2010; Miyazaki et al., 2015); this kind of production is expected to make actual mass production of artificial structured elements viable by the NIL in the near future.

1.7 Summary

In this chapter, we first addressed the fundamental optical properties of metals (Au, Ag, Al, Cu, Pt, Ti, and Ni) in terms of permittivity and described the basics of SPPs (i.e., single MI interface modes) in Sections 1.1–1.3.

In Section 1.4, the history of plasmon-related studies, that is, SPPs, nanotechnology-based plasmonics, and MMs, was briefly overviewed. Also, the recent trends such as metasurfaces were addressed.

In Section 1.5, the numerical methods applicable to plasmonic resonators were described. In particular, the RCWA+S-matrix method was elaborated. The FDTD and FEM methods were also referred to.

In Section 1.6, the nanofabrication techniques to produce plasmonic resonators were noted. As the high-precision techniques, EBL and NIL were described.

This chapter has provided the minimum and practical knowledge to launch the studies on plasmonic resonators. The physics at resonances is described in the next chapter.

References

Aieta, F., Genevet, P., Kats, M. A., Yu, N., Blanchard, R., Gaburro, Z., and Capasso, F. (2012). Aberration-free ultrathin flat lenses and axicons at telecom wavelengths based on plasmonic metasurfaces, *Nano Lett.* **12**, 4932–4936.

Alexander, R. W., Kovener, G. S., and Bell, R. J. (1974). Dispersion curves for surface electromagnetic waves with damping, *Phys. Rev. Lett.* **32**, 154–157.

Ameling, R., and Giessen, H. (2010). Cavity plasmonics: Large normal mode splitting of electric and magnetic particle plasmons induced by a photonic microcavity, *Nano Lett.* **10**, 4394–4398.

Anker, J. N., Hall, W. P., Lyandres, O., Shah, N. C., Zhao, J., and van Duyne, R. P. (2008). Biosensing with plasmonic nanosensors, *Nature Mater.* **7**, 442–453.

Arakawa, E. T., Williams, M. W., Hamm, R. N., and Ritchie, R. H. (1973). Effect of damping on surface plasmon dispersion, *Phys. Rev. Lett.* **31**, 1127–1129.

Ashcroft, N. W., and Mermin, N. D. (1976). *Solid State Physics*, international edn. (Saunders College, Orlando).

Bloembergen, N., Chang, R. K., Jha, S. S., and Lee, C. H. (1968). Optical second-harmonic generation in reflection from media with inversion symmetry, *Phys. Rev.* **174**, 813–822.

Boyd, G. T., Rasing, T., Leite, J. R. R., and Shen, Y. R. (1984). Local-field enhancement on rough surfaces of metals, semimetals, and semiconductors with the use of optical second-harmonic generation, *Phys. Rev. B* **30**, 519–526.

Brolo, A. G. (2012). Plasmonics for future biosensors, *Nature Photon.* **6**, 709–713.

Chen, C. K., de Castro, A. R. B., and Shen, Y. R. (1981). Surface-enhanced second-harmonic generation, *Phys. Rev. Lett.* **46**, 145–148.

Cho, D. J., Wu, W., Wang, F., and Shen, Y. R. (2014). Probing the plasmonic band structure of an optical metamaterial, *Phys. Rev. B* **89**, 035434.

Cho, K. (2010). *Reconstruction of Macroscopic Maxwell Equations* (Springer, Berlin).

Chou, S. Y., Krauss, P. R., and Renstorm, P. J. (1995). Imprint of sub-25 nm vias and trenches in polymers, *Appl. Phys. Lett.* **67**, 3114–3116.

Chou, S. Y., Krauss, P. R., Zhang, W., Guo, L., and Zhuang, L. (1997). Sub-10 nm imprint lithography and applications, *J. Vac. Sci. Technol. B* **15**, 2897–2904.

Depine, R. A., and Lakhtakia, A. (2004). Comment I on "resonant and antiresonant frequency dependence of the effective parameters of metamaterials", *Phys. Rev. E* **70**, 048601.

Dolling, G., Enkrich, C., Wegener, M., Soukoulis, C. M., and Linden, S. (2006). Low-loss negative-index metamaterial at telecommunication wavelengths, *Opt. Lett.* **31**, 1800–1802.

Dolling, G., Wegener, M., Soukoulis, C. M., and Linden, S. (2007). Negative-index metamaterial at 780 nm wavelength, *Opt. Lett.* **32**, 53–55.

Ebbesen, T. W., Lezec, H. J., Ghaemi, H. F., Thio, T., and Wolff, P. A. (1998). Extraordinary optical transmission through sub-wavelength hole arrays, *Nature* **391**, 667–669.

Economou, E. N. (1969). Surface plasmons in thin films, *Phys. Rev.* **182**, 539–554.

Efros, A. L. (2004). Comment II on "resonant and antiresonant frequency dependence of the effective parameters of metamaterials", *Phys. Rev. E* **70**, 048602.

Enkrich, C., Wegener, M., Linden, S., Burger, S., Zschiedrich, L., Schmidt, F., Zhou, J. F., Koschny, T., and Soukoulis, C. M. (2005). Magnetic metamaterials at telecommunication and visible frequencies, *Phys. Rev. Lett.* **95**, 203901.

Fang, N., Lee, H., Sun, C., and Zhang, X. (2005). Sub-diffraction-limited optical imaging with a silver superlens, *Science* **308**, 534–537.

Faraday, M. (1857). Experimental relations of gold (and other metals) to light, *Philos. Trans. R. Soc. Lond.* **147**, 145–181.

Fleischmann, M., Hendra, P. J., and McQuillan, A. J. (1974). Raman spectra of pyridine adsorbed at a silver electrode, *Chem. Phys. Lett.* **26**, 163–166.

Garcia-Vidal, F. J., Martin-Moreno, L., Ebbesen, T. W., and Kuipers, L. (2010). Light passing through subwavelength apertures, *Rev. Mod. Phys.* **82**, 729–787.

Genet, C., and Ebbesen, T. W. (2007). Light in tiny holes, *Nature* **445**, 39–46.

Ghaemi, H. F., Thio, T., Grupp, D. E., Ebbesen, T. W., and Lezec, H. J. (1998). Surface plasmons enhance optical transmission through subwavelength holes, *Phys. Rev. B* **58**, 6779–6782.

Granet, G., and Guizal, B. (1996). Efficient implementation of the coupled-wave method for metallic lamellar gratings in TM polarization, *J. Opt. Soc. Am. A* **13**, 1019–1023.

Ishi, T., Fujikata, J., Makita, K., Baba, T., and Ohashi, K. (2005). Si nano-photodiode with a surface plasmon antenna, *Jpn. J. Appl. Phys.* **44**, L364–L366.

Ishikawa, A., Tanaka, T., and Kawata, S. (2005). Negative magnetic permeability in the visible light region, *Phys. Rev. Lett.* **95**, 237401.

Iwanaga, M. (2010a). Polarization-selective transmission in stacked two-dimensional complementary plasmonic crystal slabs, *Appl. Phys. Lett.* **96**, 083106.

Iwanaga, M. (2010b). Subwavelength electromagnetic dynamics in stacked complementary plasmonic crystal slabs, *Opt. Express* **18**, 15389–15398.

Iwanaga, M. (2010c). Electromagnetic eigenmodes in a stacked complementary plasmonic crystal slab, *Phys. Rev. B* **82**, 155402.

Iwanaga, M. (2011). In-plane plasmonic modes of negative group velocity in perforated waveguides, *Opt. Lett.* **36**, 2504–2506.

Iwanaga, M. (2012a). *Collective Plasmonic States Emerged in Metallic Nanorod Array and Their Application* in *Nanorods* (InTech, Rijeka, Croatia), 75–92.

Iwanaga, M. (2012b). First-principle analysis for electromagnetic eigen modes in an optical metamaterial slab, *Prog. Electromagn. Res.* **132**, 129–148.

Iwanaga, M. (2012c). Photonic metamaterials: a new class of materials for manipulating light waves, *Sci. Technol. Adv. Mater.* **13**, 053002.

Iwanaga, M., Vengurlekar, A. S., Hatano, T., and Ishihara, T. (2007). Reciprocal transmittances and reflectances: An elementary proof, *Am. J. Phys.* **75**, 899–902.

Jackson, J. D. (1999). *Classical Electrodynamics*, 3rd edn. (Wiley, Hoboken, NJ).

Johnson, S. G., and Jannopoulos, J. D. (2001) *Photonic Crystals: The Road from Theory to Practice* (Kluwer Academic, Massachusetts).

Josephson, B. D. (1962). Possible new effects in superconductive tunnelling, *Phys. Lett.* **1**, 251–253.

Kadic, M., Bückmann, T., Schittny, R., and Wegener, M. (2013). Metamaterials beyond electromagnetism, *Rep. Prog. Phys.* **76**, 126501.

Kittel, C. (2004). *Introduction to Solid State Physics*, 8th edn. (Wiley, New York).

Klein, M. W., Enkrich, C., Wegener, M., Soukoulis, C. M., and Linden, S. (2006). Single-slit split-ring resonators at optical frequencies: limits of size scaling, *Opt. Lett.* **31**, 1259–1261.

Ko, D. Y. K., and Inkson, J. C. (1988). Matrix method for tunneling in heterostructures: Resonant tunneling in multilayer systems, *Phys. Rev. B* **38**, 9945–9951.

Koschny, T., Markoš, P., Smith, D. R., and Soukoulis, C. M. (2003). Resonant and antiresonant frequency dependence of the effective parameters of metamaterials, *Phys. Rev. E* **68**, 065602.

Koschny, T., Markoš, P., Smith, D. R., and Soukoulis, C. M. (2004). Reply to comments on "resonant and antiresonant frequency dependence of the effective parameters of metamaterials", *Phys. Rev. E* **70**, 048603.

Lakowicz, J. R., Ray, K., Chowdhury, M., Szmacinski, H., Fu, Y., Zhang, J., and Nowaczyk, K. (2008). Plasmon-controlled fluorescence: a new paradigm in fluorescence spectroscopy, *Analyst* **133**, 1308–1348.

Lalanne, P., and Morris, G. M. (1996). Highly improved convergence of the coupled-wave method for TM polarization, *J. Opt. Soc. Am. A* **13**, 779–784.

Lambe, J., and McCarthy, S. L. (1976). Light emission from inelastic electron tunneling, *Phys. Rev. Lett.* **37**, 923–925.

Landau, L. D., Lifshitz, E. M., and Pitaevskii, P. (1984). *Electrodynamics of Continuous Media*, 2nd edn. (Pergamon, New York).

Leonhardt, U. (2006). Optical conformal mapping, *Science* **312**, 1777–1780.

Li, L. (1996a). Formulation and comparison of two recursive matrix algorithm for modeling layered diffraction gratings, *J. Opt. Soc. Am. A* **13**, 1024–1035.

Li, L. (1996b). Use of Fourier series in the analysis of discontinuous periodic structures, *J. Opt. Soc. Am. A* **13**, 1870–1876.

Li, L. (1997). New formulation of the Fourier modal method for crossed surface-relief gratings, *J. Opt. Soc. Am. A* **14**, 2758–2767.

Linden, S., Enkrich, C., Wegener, M., Zhou, J., Koschny, T., and Soukoulis, C. M. (2004). Magnetic response of metamaterials at 100 terahertz, *Science* **306**, 1351–1353.

Lourtioz, J.-M., Benisty, H., Berger, V., Gérard, J.-M., Maystre, D., and Tchelnokov, A. (2005). *Photonic Crystals Towards Nanoscale Photonic Devices*, translated edn. (Springer, Berlin).

Maier, S. A., and Atwater, H. A. (2005). Plasmonics: Localization and guiding of electromagnetic energy in metal/dielectric structures, *J. Appl. Phys.* **98**, 011101.

Maier, S. A., Brongersma, M. L., Kik, P. G., Meltzer, S., Requicha, A. A. G. and Atwater, H. A. (2001). Plasmonics — a route to nanoscale optical devices, *Adv. Mater.* **13**, 1501–1505.

Mäkelä, T., and Haatainen, T. (2012). Roll-to-roll pilot nanoimprinting process for backlight devices, *Microelec. Eng.* **97**, 89–91.

Mäkelä, T., Haatainen, T., and Ahopelto, J. (2011). Roll-to-roll printed gratings in cellulose acetate web using novel nanoimprinting device, *Microelec. Eng.* **88**, 2045–2047.

Markoš, P., and Soukoulis, C. M. (2008). *Wave Propagation From Electrons to Photonic Crystals and Left-Handed Materials* (Princeton University Press, Princeton, NJ).

Marton, L., Simpson, J. A., Fowler, H. A., and Swanson, N. (1962). Plural scattering of 20-kev electrons in aluminum, *Phys. Rev.* **126**, 182–192.

Maxwell-Garnett, J. C. (1904). Colours in metal glasses and in metallic films, *Philos. Trans. R. Soc. Lond. A* **203**, 385–420.

Maxwell-Garnett, J. C. (1906). Colours in metal glasses, in metallic films, and in metallic solutions. II, *Philos. Trans. R. Soc. Lond. A* **205**, 237–288.

McCarthy, S. L., and Lambe, J. (1977). Enhancement of light emission from metal-insulator-metal tunnel junctions, *Appl. Phys. Lett.* **30**, 427–429.

McCarthy, S. L., and Lambe, J. (1978). Leit effect in metal-insulator-semiconductor tunnel junctions, *Appl. Phys. Lett.* **33**, 858–860.

Miyazaki, H. T., Kasaya, T., Oosato, H., Sugimoto, Y., Choi, B., Iwanaga, M. and Sakoda, K. (2015). Ultraviolet-nanoimprinted packaged metasurface thermal emitters for infrared CO_2 sensing, *Sci. Technol. Adv. Mater.* **16**, 035005.

Mock, J. J., Barbic, M., Smith, D. R., Schultz, D. A., and Schultz, S. (2002). Shape effects in plasmon resonance of individual colloidal silver nanoparticles, *J. Chem. Phys.* **116**, 6755–6759.

Murray, A., and Barnes, W. L. (2007). Plasmonic materials, *Adv. Mater.* **19**, 3771–3782.

Ofir, Y., Moran, I. W., Subramani, C., Carter, K. R., and Rotello, V. M. (2010). Nanoimprint lithography for functional three-dimensional patterns, *Adv. Mater.* **22**, 3608–3614.

Okamoto, T., H'Dhili, F., and Kawata, S. (2004). Towards plasmonic band gap laser, *Appl. Phys. Lett.* **85**, 3968–3970.

Palik, E. D. (1991). *Handbook of Optical Constants of Solids II* (Academic, San Diego, USA).

Pendry, J. B. (2000). Negative refraction makes a perfect lens, *Phys. Rev. Lett.* **85**, 3966–3969.

Pendry, J. B., Holden, A. J., Robbins, D. J., and Stewart, W. J. (1999). Magnetism from conductors and enhanced nonlinear phenomena, *IEEE Trans. Micro. Theor. Technol.* **47**, 2075–2084.

Pendry, J. B., Schurig, D., and Smith, D. R. (2006). Controlling electromagnetic fields, *Science* **312**, 1780–1782.

Potton, R. J. (2004). Reciprocity in optics, *Rep. Prog. Phys.* **67**, 717–754.

Powell, D. A., Shadrivov, I. V., and Kivshar, Y. S. (2009). Nonlinear electric metamaterials, *Appl. Phys. Lett.* **95**, 084102.

Raether, H. (1988). *Surface Plasmons on Smooth and Rough Surfaces and on Gratings* (Springer-Verlag, Berlin).

Rakić, A. D., Djurušić, A. B., Elazar, J. M., and Majewski, M. L. (1998). Optical properties of metallic films for vertical-cavity optoelectronic devices, *Appl. Opt.* **37**, 5271–5283.

Rodrigo, S. G. (2012). *Optical Properties of Nanostructured Metallic Systems* (Springer, Berlin).

Sakoda, K. (2005). *Optical Properties of Photonic Crystals*, 2nd edn. (Springer, Berlin).

Schurig, D., Mock, J. J., Justice, B. J., Cummer, S. A., Pendry, J. B., Starr, A. F., and Smith, D. R. (2006). Metamaterial electromagnetic cloak at microwave frequencies, *Science* **314**, 977–980.

Shalaev, V. M., Cai, W., Chettiar, U. K., Yuan, H.-K., Sarychev, A. K., Drachev, V. P., and Kildishev, A. V. (2005). Negative index of refraction in optical metamaterials, *Opt. Lett.* **30**, 3356–3358.

Shelby, R. A., Smith, D. R., and Schultz, S. (2001). Experimental verification of a negative index of refraction, *Science* **292**, 77–79.

Smith, D. R., Schultz, S., Markoš, P., and Soukoulis, C. M. (2002). Determination of effective permittivity and permeability of metamaterials from reflection and transmission coefficient, *Phys. Rev. B* **65**, 195401.

Swihart, J. C. (1961). Field solution for a thin-film superconducting strip transmission line, *J. Appl. Phys.* **32**, 461–469.

Tao, H., Strikwerda, A. C., Fan, K., Padilla, W. J., Zhang, X., and Averitt, R. D. (2009). Reconfigurable terahertz metamaterials, *Phys. Rev. Lett.* **103**, 147401.

Thio, T., Pellerin, K. M., Linke, R. A., Lezec, H. J., and Ebbesen, T. W. (2001). Enhanced light transmission through a single subwavelength aperture, *Opt. Lett.* **26**, 1972–1974.

Tonks, L., and Langmuir, I. (1929). Oscillations in ionized gases, *Phys. Rev.* **33**, 195–210.

Valentine, J., Zhang, S., Zentgraf, T., Ulin-Avila, E., Genov, D. A., Bartal, G., and Zhang, X. (2008). Three-dimensional optical metamaterial with a negative refractive index, *Nature* **455**, 376–379.

Vesborg, P. C. K., Seger, B., and Chorkendorff, I. (2015). Recent development in hydrogen evolution reaction catalysts and their practical implementation, *J. Phys. Chem. Lett.* **6**, 951–957.

Volakis, J. L., Chatterjee, A., and Kempel, L. C. (1994). Review of the finite-element method for three-dimensional electromagnetic scattering, *J. Opt. Soc. Am. A* **11**, 1422–1433.

Wang, C., Zhang, Q., Song, Y., and Chou, S. Y. (2014). Plasmonic bar-coupled dots-on-pillar cavity antenna with dual resonances for infrared absorption and sensing: Performance and nanoimprint fabrication, *ACS Nano* **8**, 2618–2624.

Wooten, F. (1972). *Optical Properties of Solids* (Academic Press, New York).

Yee, K. S. (1966). Numerical solution of initial boundary value problems involving Maxwell's equations in isotropic media, *IEEE Trans. Antennas Propagat.* **14**, 302–307.

Yen, T. J., Padilla, W. J., Fang, N., Vier, D. C., Smith, D. R., Pendry, J. B., Basov, D. N., and Zhang, X. (2004). Terahertz magnetic response from artificial materials, *Science* **303**, 1494–1496.

Yu, N., Genevet, P., Kats, M. A., Aieta, F., Tetienne, J.-P., Capasso, F., and Gaburro, Z. (2011). Light propagation with phase discontinuities: Generalized laws of reflection and refraction, *Science* **334**, 333–337.

Zhang, J., MacDonald, K. F., and Zheludev, N. I. (2012). Optical gecko toe: Optically controlled attractive near-field forces between plasmonic metamaterials and dielectric or metal surfaces, *Phys. Rev. B* **85**, 205123.

Zhang, S., Fan, W., Panoiu, N. C., Malloy, K. J., Osgood, R. M., and Brueck, S. R. J. (2005). Experimental demonstration of near-infrared negative-index metamaterials, *Phys. Rev. Lett.* **95**, 137404.

Zhang, S., Xia, C., and Fang, N. (2011). Broadband acoustic cloak for ultrasound waves, *Phys. Rev. Lett.* **106**, 024301.

Zheludev, N. I., and Kivshar, Y. S. (2012). From metamaterials to metadevices, *Nature Mater.* **11**, 917–924.

Zheng, G., Mühlenbernd, H., Kenney, M., Li, G., Zentgraf, T., and Zhang, S. (2015). Metasurface holograms reaching 80% efficiency, *Nature Nanotech.* **10**, 308–312.

Zhou, J., Koschny, T., Kafesaki, M., Economou, E. N., Pendry, J. B., and Soukoulis, C. M. (2005). Saturation of the magnetic response of split-ring resonators at optical frequencies, *Phys. Rev. Lett.* **95**, 223902.

Chapter 2

Response Function Theory

General formalism for linear responses is one of the most established theoretical framework in physics, given by using response function χ:

$$\mathbf{P}(\mathbf{r}, t) = \int d\mathbf{r}' \int_{-\infty}^{t} dt' \chi(\mathbf{r}, \mathbf{r}', t, t')\mathbf{F}_{\text{ext}}(\mathbf{r}', t') \tag{2.1}$$

where $\mathbf{P}$ denotes response field (or polarization), which is generally vector, and $\mathbf{F}_{\text{ext}}$ applied external field. When $\mathbf{F}$ is electric field, $\mathbf{P}$ is electric polarization; when $\mathbf{F}$ is magnetic field, $\mathbf{P}$ is magnetic polarization (Wooten, 1972; Landau et al., 1984).

Figure 2.1 schematically depicts response of materials or artificial structures including plasmonic resonators for the applied

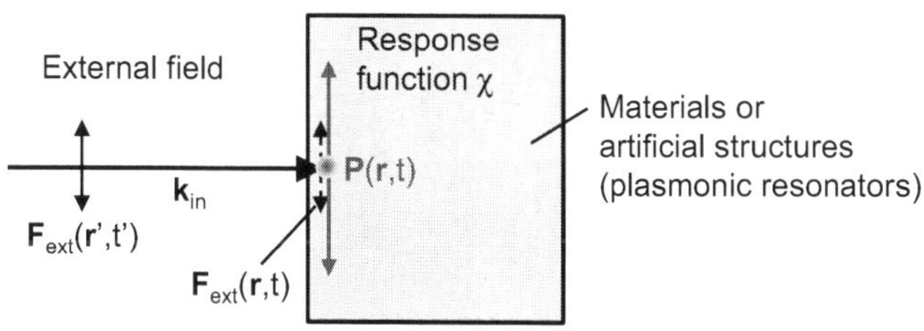

Figure 2.1 Schematic of response function χ, defined in Eq. 2.1.

Plasmonic Resonators: Fundamentals, Advances, and Applications
Masanobu Iwanaga

ISBN 978-981-4745-06-2 (Hardcover), 978-981-4745-07-9 (eBook)
www.panstanford.com

external field $\mathbf{F}_{\text{ext}}$. The response function χ works as expressed in Eq. 2.1.

Here we modify Eq. 2.1 in a different representation as follows (Landau et al., 1984). Generally, we can prepare monochromatic external field such that

$$\mathbf{F}_{\text{ext}}(\mathbf{r}', t') = \mathbf{F}_{\text{ext}}(\mathbf{k}, \omega) \exp(i\mathbf{k} \cdot \mathbf{r}' - i\omega t') \tag{2.2}$$

where $\mathbf{k}$ denotes wavevector and ω angular frequency. It is usually reasonable to assume that $\chi(\mathbf{r}, \mathbf{r}', t, t')$ is equal to $\chi(\mathbf{r}, \mathbf{r}', t-t')$; then, Eq. 2.1 is slightly modified such that

$$\mathbf{P}(\mathbf{r}, t) = \int d\mathbf{r}' \int_0^\infty ds \chi(\mathbf{r}, \mathbf{r}', s) \mathbf{F}_{\text{ext}}(\mathbf{r}', t - s). \tag{2.3}$$

When we define $\chi(\mathbf{r}, \mathbf{r}', \omega)$ by

$$\chi(\mathbf{r}, \mathbf{r}', \omega) = \int_0^\infty ds \chi(\mathbf{r}, \mathbf{r}', s) \exp(i\omega s), \tag{2.4}$$

a standard Fourier transformation from time t to frequency ω results in the following relation

$$\mathbf{P}(\mathbf{r}, \omega) = \int d\mathbf{r}' \chi(\mathbf{r}, \mathbf{r}', \omega) \mathbf{F}_{\text{ext}}(\mathbf{k}, \omega) \exp(i\mathbf{k} \cdot \mathbf{r}'). \tag{2.5}$$

If we further assume that $\chi(\mathbf{r}, \mathbf{r}', \omega)$ is equal to $\chi(\mathbf{r} - \mathbf{r}', \omega)$, we can carry out a procedure on $\mathbf{r}'$ and $\mathbf{r}$, similar to Eqs. 2.3–2.5, and reach the $(\mathbf{k}, \omega)$-representation of Eq. 2.1 such that

$$\mathbf{P}(\mathbf{k}, \omega) = \chi(\mathbf{k}, \omega) \mathbf{F}_{\text{ext}}(\mathbf{k}, \omega). \tag{2.6}$$

Equation 2.6 means that nonlocal response of $\chi(\mathbf{r}, \mathbf{r}', \omega)$ (that is, $\chi(\mathbf{r}, \mathbf{r}', \omega) \neq 0$ for $\mathbf{r} \neq \mathbf{r}'$) in Eq. 2.1 is equivalent to spatial dispersion of $\chi(\mathbf{k}, \omega)$ under the usual assumptions. Therefore, the spatial dispersion of χ, i.e., $\mathbf{k}$-dependence of χ, is understood as evidence of the nonlocal response.

In this chapter, we outlook response functions in classical regime (Section 2.1) and in quantum mechanical regime (Section 2.2). On the basis of the response functions, spectral shapes near resonances of plasmonic resonators are described as Fano resonances (Section 2.3). Further, generalized response functions for nanostructures, which include spatial dispersion, are addressed (Section 2.4).

2.1 Classical Model for Response Function

When the response function χ in Eq. 2.1 is restricted to local response such that

$$\chi(\mathbf{r}, \mathbf{r}', t, t') = \chi(\mathbf{r}', t')\delta(\mathbf{r} - \mathbf{r}')\delta(t - t'), \tag{2.7}$$

Equation 2.1 is reduced to the following local-response representation:

$$\mathbf{P}(\mathbf{r}, t) = \chi(\mathbf{r}, t)\mathbf{F}_{\text{ext}}(\mathbf{r}, t). \tag{2.8}$$

We note that the local response does not lead the spatial dispersion of χ, shown in Eq. 2.6.

In a general form, electric polarization vector **P** contributes to electric density flux **D** in the following:

$$\mathbf{D} = \varepsilon_0\mathbf{E} + \mathbf{P} \tag{2.9}$$

$$= \varepsilon_0(1 + \chi)\mathbf{E}. \tag{2.10}$$

To reach Eq. 2.10, we used Eq. 2.8. Recalling the constitutive equation Eq. 1.11, relative permittivity ε in linear response regime is expressed as

$$\varepsilon = 1 + \chi. \tag{2.11}$$

Thus, relative permittivity is explicitly connected to the electric(E)-field response function χ, which is usually called susceptibility.

Let us consider the case of E-field application ($F_{\text{ext}} = E_{\text{ext}}$). Classically, relative permittivity ε is derived from Lorentz oscillator model that comprises a damping-associated forced oscillator model as follows (Jackson, 1999; Kobayashi, 1997):

$$m\ddot{x} + m\gamma\dot{x} + m\omega_0^2 x = -eE_{\text{ext}} \tag{2.12}$$

where m is mass of the oscillator, γ denotes damping factor, ω_0 is resonant frequency of the oscillator, and e is charge of the oscillator ($e>0$). When the position $x(t)$ of the oscillator is driven by the external field $E_{\text{ext}}(t) = E_{\text{ext}}\exp(-i\omega t)$, the position $x(t)$ is expressed as

$$x(t) = x\exp(-i\omega t),$$

where the amplitudes E_{ext} and x are independent of t. Equation 2.12 is modified to the displacement of the oscillator such as

$$x = \frac{-eE_{\text{ext}}}{m(\omega_0^2 - \omega^2 - i\gamma\omega)}. \tag{2.13}$$

Polarization P is induced by the E_{ext} such that

$$P = N\sqrt{f}e|x|$$

where $\sqrt{f}$ is a coefficient in this model. Using Eqs. 2.10 and 2.11, we reach the relative permittivity ε in the Lorentz model:

$$\varepsilon(\omega) = 1 + \frac{f\omega_p^2}{\omega_0^2 - \omega^2 - i\gamma\omega}. \tag{2.14}$$

This function is not Lorentzian function though the spectral shape is similar. The complex $\varepsilon(\omega)$ is separated into the real and imaginary parts:

$$\text{Re}[\varepsilon(\omega)] = 1 + \frac{f\omega_p^2(\omega_0^2 - \omega^2)}{(\omega_0^2 - \omega^2)^2 + \gamma^2\omega^2} \tag{2.15}$$

$$\text{Im}[\varepsilon(\omega)] = \frac{f\omega_p^2\gamma\omega}{(\omega_0^2 - \omega^2)^2 + \gamma^2\omega^2}. \tag{2.16}$$

From Eq. 2.11, susceptibility χ has Lorentzian form such that

$$\chi(\omega) = \frac{f\omega_p^2}{\omega_0^2 - \omega^2 - i\gamma\omega}. \tag{2.17}$$

Figure 2.2 shows illustrative results on the relative permittivity in Eq. 2.14, equivalent to Eqs. 2.15 and 2.16. Figure 2.2a is a typical distribution of the relative permittivity in Eq. 2.14; the parameters were set to be a realistic set such as $f = 0.01$, $\hbar\omega_p = 10$ eV, $\hbar\omega_0 = 1.5$ eV, and $\hbar\gamma = 0.1$ eV. Re(ε) takes positive and negative values around the resonant energy $\hbar\omega_0$, indicated by arrow. Im(ε) has the peak at $\hbar\omega_0$, which means that ω_0 is resonant frequency of the damping oscillator driven by the external field in Eq. 2.12; the line width is determined by the damping factor γ.

Figure 2.2b shows reflectance (R) spectrum of bulk material of the ε. The R at the normal incidence was evaluated in the configuration of Fig. 1.4, using Eq. 1.22 in Section 1.2; the incident layer was assumed to be air and the relative permeability was set to unity. The R spectrum directly represents the properties of $\varepsilon(\omega)$; at

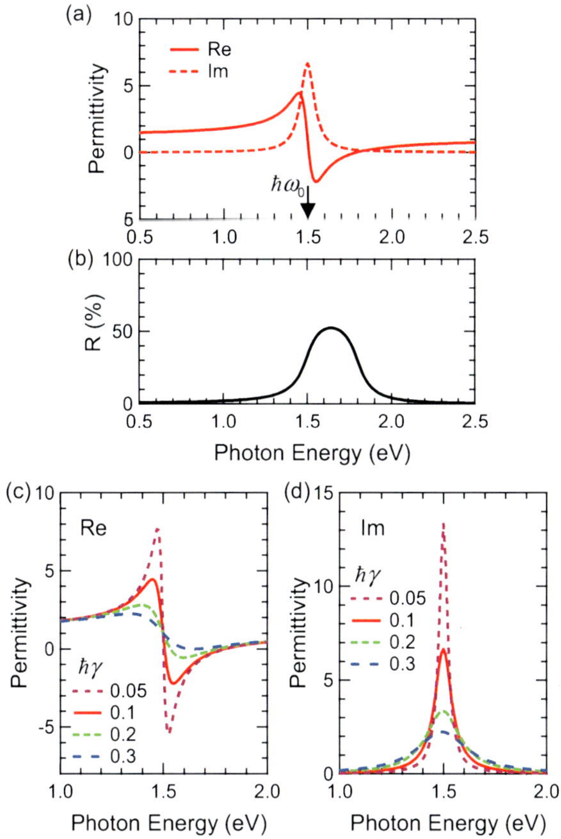

Figure 2.2 (a) Relative permittivity $\varepsilon(\omega)$ in the classical model expressed by Eq. 2.14 with $f = 0.01$, $\hbar\omega_p = 10$ eV, $\hbar\omega_0 = 1.5$ eV (arrow), and $\hbar\gamma = 0.1$ eV. (b) Calculated R spectrum of semi-infinitely thick material of $\varepsilon(\omega)$. (c) and (d) The real and imaginary parts of the relative permittivity with different γ, respectively; $\hbar\gamma = 0.05$, 0.1, 0.2, and 0.3 eV.

the off-resonance frequency ranges (i.e., $\hbar\omega \leq 0.8$ and $\hbar\omega \geq 2.2$ eV), the material of ε is transparent like air because of $\varepsilon \approx 1$ whereas, at the resonant range (i.e., $1.0 < \hbar\omega < 2.0$ eV), the ε varies at a significant amount and results in the large change of R. In particular, note that the peak of R spectra is located at the higher energy than the resonant energy $\hbar\omega_0$, and that the R band of R > 25% almost corresponds to the range of $\text{Re}[\varepsilon(\omega)] < 0$.

Figures 2.2c and 2.2d present the real and imaginary parts of $\varepsilon(\omega)$ with different damping factors γ, respectively. The case of $\hbar\gamma = 0.05$ (eV) is shown with purple dotted line, $\hbar\gamma = 0.1$ with red line, $\hbar\gamma = 0.2$ with green broken line, and $\hbar\gamma = 0.3$ with blue dashed line. The case of $\hbar\gamma = 0.1$ is in common with Fig. 2.2a. As γ becomes larger, the peak value reduces. We mention that it is always satisfied such that

$$\mathrm{Re}[\varepsilon(\omega_0)] = 1 \tag{2.18}$$

in Eq. 2.14 while the imaginary part,

$$\mathrm{Im}[\varepsilon(\omega_0)] = \frac{f\omega_p^2}{\gamma\omega_0}, \tag{2.19}$$

indicates that the peak value is inversely proportional to ω_0 and γ. Indeed, relative permittivity takes large values more than 100 at lower frequencies such as microwaves. Besides, a relation of $\omega_0 \gg \gamma/2$ is often satisfied and leads an approximate equation such that

$$\mathrm{Re}[\varepsilon(\omega_0 - \gamma/2)] \approx 1 + \frac{1}{2}\frac{f\omega_p^2}{\gamma\omega_0} \tag{2.20}$$

and then the $\mathrm{Re}(\varepsilon)$ also takes large values near the resonant frequency ω_0. It is verified that the relations in Eqs. 2.18–2.20 hold in Fig. 2.2.

2.2 Quantum Mechanical Description for Response Function

Modern physics is constructed based on quantum mechanics; it is therefore preferred to provide a quantum mechanical description for response function, which has been derived from the classical model in Section 2.1. Here, formulation for response function in the quantum mechanical regime is described. Historically, the general formulation (Fano, 1956) and a concrete application to excitons[a] (Hopfield, 1958) were reported for in 1950s. As a result, an equivalent dielectric function to the Lorentz model is derived, making it possible for the classical model to survive.

---

[a] Excitons are elementary electronic excitations in insulators and semiconductors that have bandgap between the valence and conduction bands. A pair of electron and hole is excited below the bandgap energy and forms a bound state via the attractive Coulomb force.

Theoretical setting Let us start setting a general interaction system composed of matter including discrete-level oscillators and of electromagnetic (EM) fields (Fano, 1956). The matter can be dense atoms, solids, or liquids. The total Hamiltonian of the system $\mathcal{H}$ is express as

$$\mathcal{H} = \mathcal{H}_{\mathrm{m}} + \mathcal{H}_{\mathrm{EM}} + \mathcal{H}_{\mathrm{int}} \tag{2.21}$$

where H_{m} is Hamiltonian of matter, H_{EM} Hamiltonian of EM fields, and H_{int} Hamiltonian of the matter–EM-field interaction. These terms are written as

$$\mathcal{H}_{\mathrm{m}} = \sum_{n,\mathbf{k},s} \frac{\hbar\omega_n}{2}\left(P_{n,\mathbf{k},s}P^*_{n,\mathbf{k},s} + \omega_n^2 Q_{n,\mathbf{k},s}Q^*_{n,\mathbf{k},s}\right) \tag{2.22}$$

$$\mathcal{H}_{\mathrm{EM}} = \sum_{\mathbf{k},s} \frac{1}{2}\left\{p_{\mathbf{k},s}p^*_{\mathbf{k},s} + c^2k^2(1-\delta_{s,1})q_{\mathbf{k},s}q^*_{\mathbf{k},s}\right\} \tag{2.23}$$

$$\mathcal{H}_{\mathrm{int}} = \sum_{n} \omega_p\sqrt{f_n}\,q_{\mathbf{k},s}P_{n,\mathbf{k},s} \tag{2.24}$$

where P and Q are operators for matter that has discrete energy levels $\hbar\omega_n$, p and q are operators for EM fields, $\mathbf{k}$ is wavenumber vector ($k = |\mathbf{k}|$), s runs from 1 to 3, ω_p is the plasma frequency defined by Eq. 1.3, and f_n is oscillator strength in the dipole approximation. Note that Eq. 2.23 assumes the situation in absence of genuine current.

Eigenvalue equation and dielectric function The eigenvalue equations of the Hamiltonian in Eq. 2.21 are obtained in the following.

$$\begin{aligned}\frac{\partial\mathcal{H}}{\partial q^*_{\mathbf{k},s}} &= \left\{c^2k^2(1-\delta_{s,1}) + \omega_p^2\right\}q_{\mathbf{k},s} + \sum_n \omega_p\sqrt{f_n}P^*_{n,\mathbf{k},s}\\ &= \omega^2 q_{\mathbf{k},s}\end{aligned} \tag{2.25}$$

$$\begin{aligned}\frac{\partial\mathcal{H}}{\partial P_{n,\mathbf{k},s}} &= \omega_p\sqrt{f_n}\,q_{\mathbf{k},s} + \omega_n P^*_{n,\mathbf{k},s}\\ &= \omega^2 P^*_{n,\mathbf{k},s}/\omega_n.\end{aligned} \tag{2.26}$$

From Eq. 2.26, we obtain

$$P^*_{n,\mathbf{k},s} = \frac{\sqrt{f_n}\,\omega_p\omega_n}{\omega^2-\omega_n^2}q_{\mathbf{k},s}. \tag{2.27}$$

Substituting Eq. 2.27 into Eq. 2.25, we finally reach dispersion equation in the interacting system described in Eq. 2.21:

$$\frac{c^2k^2}{\omega^2}(1-\delta_{s,1}) = 1 + \sum_n \frac{f_n\omega_p^2}{\omega_n^2 - \omega^2}. \tag{2.28}$$

We here used a relation of $\sum_n f_n = 1$, which is called sum rule. Equation 2.28 manifests the basic properties of EM waves in the interacting system as follows

(i) The dispersion equation for the transverse EM waves represents the relation (or constraint) between wavenumber k and frequency ω. The dispersion diagrams are usually plotted in the (k, ω) plane. For example, dispersion diagrams of the SPPs at flat Ag–air interface were shown in Figs. 1.8 and 1.9.
(ii) The right-hand side of Eq. 2.28 is the discrete sum of the relative permittivity of the Lorentz oscillators assuming the damping factor $\gamma = 0$ (Eq. 2.14).
(iii) The left-hand side indicates that only transverse wave contributes to the dispersion equation. The longitudinal wave component, which is represented with $s = 1$, is explicitly excluded from the dispersion equation.

Equation 2.28 is purely real-valued. To incorporate actual responses, it is needed to extend the equation to complex-valued one, which is carried out by making ω_n to be complex-valued. Concretely, $\omega_n^2 - \omega^2$ is replaced with $(\omega_n - \omega)(\omega_n^* + \omega)$ (Fano, 1956). We define a function ε by

$$\varepsilon(\omega) = 1 + \sum_n \frac{f_n\omega_p^2}{|\omega_n|^2 - \omega^2 - i\gamma_n\omega} \tag{2.29}$$

where we set $\gamma_n = 2\,\mathrm{Im}(\omega_n)$. Equation 2.28 is written as

$$c^2k^2(1-\delta_{s,1}) = \omega^2\varepsilon(\omega). \tag{2.30}$$

Equation 2.30 is a similar expression to the classical dispersion equation of media with relative permittivity ε. Thus, it is shown that Eq. 2.29 gives relative permittivity in the quantum mechanics. Note that, in general, the terms ω_n and f_n possibly depend on $\mathbf{k}$, so that $\varepsilon(\omega)$ also depends on $\mathbf{k}$. The spatial dispersions of single-crystal metal and plasmonic resonators are addressed in Chapter 4.

As for the dielectric function (i.e., relative permittivity), we have confirmed the followings.

(i) Equation 2.29 is dielectric function derived in the quantum mechanical framework.
(ii) The classical Lorentz model in Section 2.1 is safely used to provide an intuitive description for dielectric and response functions.
(iii) The dielectric function $\varepsilon(\omega)$ can be dependent on wavenumber vector $\mathbf{k}$. The spatial dispersion of single-crystal metal is addressed in Section 4.1.

Longitudinal excitations Apart from the transverse excited states described so far, longitudinal excitations are also possible (Fano, 1956). We here refer to implications from the longitudinal component.

When we consider a situation that an electron of finite kinetic energy goes in medium of dielectric function $\varepsilon(\omega)$, the loss of the kinetic energy W_{loss} is expressed such that

$$W_{\text{loss}} \propto \int \text{Im}\left(\frac{-1}{\varepsilon(\omega)}\right) d\omega. \tag{2.31}$$

As shown in Fig. 2.2d, when a peak of $\text{Im}[\varepsilon(\omega)]$ contributes at a frequency range, the W_{loss} takes a peak that can be observed in electron energy-loss spectrum (EELS) experiment (Fig. 1.3). Equation 2.31 roughly describes the principles of EELS though we here do not provide extensive descriptions for the energy-loss function. Instead, we note implications from the longitudinal excitations through the energy loss W_{loss}.

(i) The longitudinal excitations are free from constraint between wavenumber k and frequency ω; accordingly, Eq. 2.31 is often referred to as inelastic loss. This property is distinct from the transverse excitations restricted by the dispersion equation (Eq. 2.28). As a result, the longitudinal excitations are, in principle, induced at any frequency and can form continuous band.
(ii) It is also obvious from Eq. 2.31 that the efficiency of the longitudinal excitations can be significantly enhanced at the resonant frequency of $\varepsilon(\omega)$. We remark that longitudinal excitations are not negligible in artificial nanostructures such as plasmonic resonators. Note also that the longitudinal and transverse excitations are not independent but are induced simultaneously.

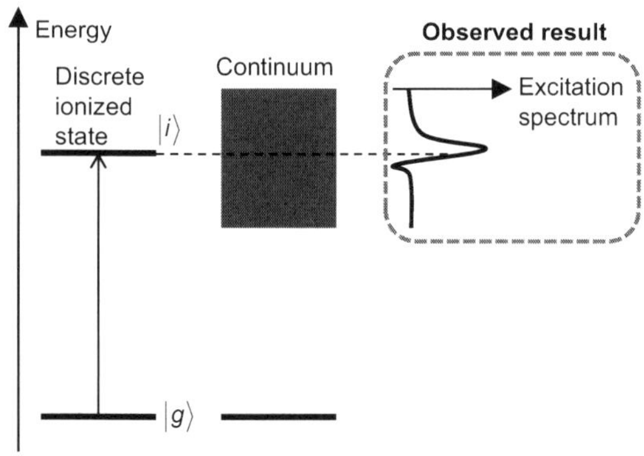

Figure 2.3 Concept of Fano resonance, originally introduced to describe the interaction of an excited discrete ionized state $|i\rangle$ of He atoms with continuum of electron kinetic energy (Fano, 1961). The resultant excitation spectrum took an asymmetric shape.

2.3 Spectral Theory

2.3.1 *Fano Resonance*

Let us here describe spectral shape theory by Fano (Fano, 1961), which was initially motivated to describe excitation spectrum of ionized states of He gas at a vacuum UV range. Figure 2.3 shows concept of the original Fano resonance. Due to the interaction of the discrete state and continuum of electron kinetic energy, the observed spectrum exhibited an asymmetric shape.

As for the theoretical framework, we here assume that a discrete electronic system and a radiation continuum (for example, EM fields) coexist in a system and the interaction takes place due to the configuration.

Theoretical setting The Hamiltonian $\mathcal{H}$ has eigenstates $|\Phi\rangle$ $(= |\phi; \psi\rangle)$ where ϕ denotes a discrete electronic states and ψ represents a continuum. Let us start explicit expressions of the submatrix for ϕ and ψ (Fano, 1961).

$$\langle\phi|\mathcal{H}|\phi\rangle = E_\phi \tag{2.32}$$

$$\langle\psi_{E'}|\mathcal{H}|\phi\rangle = V_{E'} \tag{2.33}$$

$$\langle\psi_{E'}|\mathcal{H}|\psi_{E''}\rangle = E'\delta(E' - E'') \tag{2.34}$$

Equation 2.34 assumes that the submatrix is diagonalized in advance for the subset of $\psi_{E'}$. Equations 2.32–2.34 are symbolically expressed in a matrix form for the base $|\Phi\rangle$:

$$\begin{pmatrix} E_\phi & V_{E'} \\ V_{E'}^* & [E'] \end{pmatrix}. \tag{2.35}$$

Here is called energy matrix where $[\cdot]$ denotes diagonal matrix.

Eigenvector $|\Phi_E\rangle$ for energy E is generally written as

$$|\Phi_E\rangle = a(E)|\phi\rangle + \int dE' b_{E'}(E)|\psi_{E'}\rangle. \tag{2.36}$$

From $\langle\phi|\mathcal{H}|\Phi_E\rangle$,

$$E_\phi a(E) + \int dE' V_{E'} b_{E'}(E) = Ea(E). \tag{2.37}$$

From $\langle\psi_{E'}|\mathcal{H}|\Phi_E\rangle$,

$$V_{E'}a(E) + E'b_{E'}(E) = Eb_{E'}(E). \tag{2.38}$$

Following Dirac's procedure (Dirac, 1927), Eq. 2.38 leads an expression such that

$$b_{E'}(E) = \left\{\frac{1}{E - E'} + z(E)\delta(E - E')\right\} V_{E'}a(E). \tag{2.39}$$

The newly introduced $z(E)$ is determined as follows. Substituting Eq. 2.39 into Eq. 2.37 and eliminating the factor $a(E)$, one has

$$E_\phi + \mathcal{P}\int dE' \frac{|V_{E'}|^2}{E - E'} + z(E)|V_E|^2 = E \tag{2.40}$$

where the integral takes the principal value $\mathcal{P}$. Thus, one reaches

$$z(E) = \frac{E - E_\phi - \mathcal{P}\int |V_{E'}|^2/(E - E')dE'}{|V_E|^2}. \tag{2.41}$$

Using Eq. 2.41, after modification of an equation for $\langle\Phi_{E'}|\Phi_E\rangle$, we can finally write down the amplitude of the discrete state.

$$\begin{aligned} |a(E)|^2 &= \frac{1}{\pi^2|V_E|^2[1 + \{z(E)/\pi\}^2]} \\ &= \frac{|V_E|^2}{\pi^2|V_E|^4 + \{E - E_\phi - \mathcal{P}\int |V_{E'}|^2/(E - E')dE'\}^2} \end{aligned} \tag{2.42}$$

Equation 2.42 means that the amplitude $|a(E)|^2$ has a profile of full width at the half maximum of $\pi^{-2}|V_E|^{-2}$, implying that, as the configuration interaction becomes stronger (i.e., $|V_E|$ becomes larger), the $|a(E)|^2$ distribution becomes broader or more *diluted* than the original distribution of the discrete state ϕ without the configuration interaction.

A factor $z(E)/\pi$ has a physical meaning that it is the ratio of energy around the autoionized state to the half width of the discrete state $a(E)$ in the present configuration interaction. Therefore, the $z(E)/\pi$ is a scaled energy variable and we set

$$e = \frac{z(E)}{\pi}.$$

Next, a parameter q is introduced, which represents a ratio of transition matrix elements such that

$$q = \frac{\langle\phi|T|i\rangle + \mathcal{P}\int dE'\langle\phi|H|\psi_{E'}\rangle\langle\psi_{E'}|T|i\rangle/(E - E')}{\pi\langle\phi|H|\psi_E\rangle\langle\psi_E|T|i\rangle} \tag{2.43}$$

where $|i\rangle$ is initial state and T denotes a transition operator.[b] Using Eq. 2.43, ratio of transition probability into a coupled state to transition probability into the original continuum $|\psi_E\rangle$ is expressed as

$$\frac{|\langle\Phi_E|T|i\rangle|^2}{|\langle\psi_E|T|i\rangle|^2} = \frac{(q+e)^2}{1+e^2}. \tag{2.44}$$

Equation 2.44 is usually called Fano-type line shape.

Figure 2.4 shows a series of Fano-type resonance variations. Varying the parameter q in Eq. 2.44, representative spectra were plotted. Figure 2.4a shows the cases of $q = 0$ (red solid curve), 0.5 (black dashed curve), and 1 (black solid curve). Figure 2.4b shows the case of $q \gg 1$ (blue solid curve). The Fano-type spectra have the following implications.

(i) When $q = 0$, the transition into the discrete state $|\phi\rangle$ does not take place; therefore, the value of $(q+e)^2/(1+e^2)$ goes down to 0 at $e = 0$, meaning that only transition into the continuum $|\psi_E\rangle$ takes place. In other words, the configuration interaction does not appear.

---

[b] In most cases, electric dipole transition is analyzed. However, any specific constraint for the details is not assumed here, so that other transitions such as magnetic dipole transition are also allowed in this formalism.

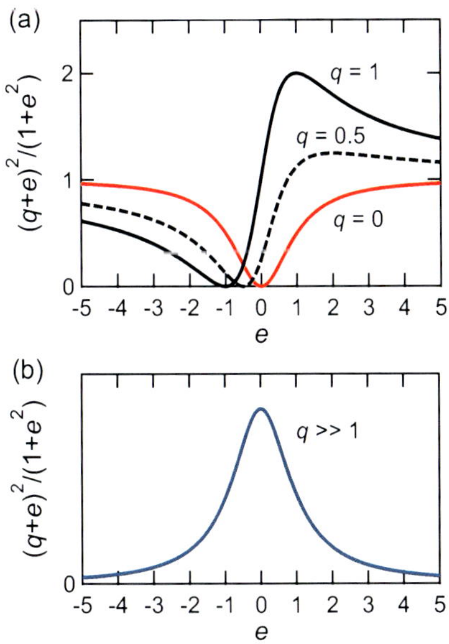

Figure 2.4 Fano-type resonance variation, derived from Eq. 2.44. (a) $q = 0$, 0.5, and 1, shown with red solid, black dashed, and black solid curves, respectively. (b) $q \gg 1$ (blue solid curve).

(ii) When $q \sim 1$, the configuration interaction exhibits a peculiar property. The spectra in Fig. 2.4 are asymmetric for the scaled energy e and are observed in optical spectra such as extinction and R spectra, as described later (see Section 2.3.2).

(iii) When $q \gg 1$, the configuration interaction becomes dominant to the non-interaction transition into the continuum. The spectrum asymptotically transforms into a symmetric shape for e such as

$$\frac{(q+e)^2}{1+e^2} \rightarrow \frac{q^2}{1+e^2}$$

and becomes a convex shape in contrast to the concave spectrum at $e = 0$, suggesting that the optical spectra exhibits qualitative changes for the case of $q = 0$.

Analyses of Fano resonances The formulation of the Fano resonance is general and applicable to various phenomena. Indeed,

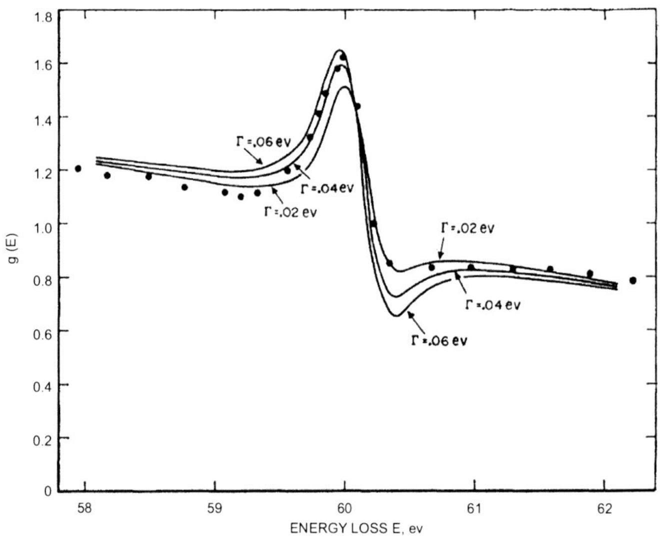

Figure 2.5 Fano resonance of a discrete ionized state of He atom and continuous electron kinetic energy. Reprinted with permission from Fano (1961). Copyright 1961 by the American Physical Society.

there have been so many experimental data exhibiting Fano-type resonances (Miroshnichenko et al., 2010; Luk'yanchuk et al., 2010).

Here we introduce two examples analyzing Fano-type resonances, which were observed in distinct configurations. Still, the underlying physics is in common, described by the present theory of configuration interaction.

Figure 2.5 shows EELS, representing resonance resulting from the configuration interaction of He atoms and an electron beam in a gas chamber. The 500 eV electron beam was absorbed and/or scattered by the individual He atoms in inelastic ways. The energy loss of the beam was measured in a monochromator. A transition in He atom around 60 eV is ascribed to the $2s2p^1P$ ionized state of He atom.

To fit the experimental data (dots) in Fig. 2.5, the ratio of transition probability in Eq. 2.44 is modified to include line width Γ such as

$$e = \frac{E - E_{\phi} - F}{\Gamma/2}. \tag{2.45}$$

where we set $F = \mathcal{P}\int dE'|V_{E'}|^2/(E - E')$. The Γ is a fitting parameter for the measured data. A function $f(E)$ is defined to consider the line slope coming from background such that

$$f(E) = 1 - 0.1e + \frac{(q+e)^2}{1+e^2} \tag{2.46}$$

where e is defined in Eq. 2.45. In the actual fitting (Fano, 1961), a convolution function $g(E)$ made of the function f was used to reproduce the data as follows.

$$g(E) = \int_{-\infty}^{\infty} \frac{1}{0.2\sqrt{\pi}} \exp\left\{-\frac{(E-E')^2}{(0.2)^2}\right\} f(E')dE' \tag{2.47}$$

In Fig. 2.5, fitted curves evaluated by Eq. 2.47 from the $\Gamma = 0.02$ to 0.06 eV are shown with solid curves. The best fitting was found at $\Gamma = 0.04$ eV. The shape was reproduced quite well. Thus, a resonant state in the system of He atom and electron beam was described in a reliable manner. This result indicates that the ionized state in a He atom is modified from ideal Lorentzian shape to an asymmetric shape due to the configuration interaction. Note that the vertical scale in Fig. 2.5 was adjusted to fit the function $g(E)$.

2.3.2 *Spectral Analysis for Fano Resonances*

Fano resonance including plasmonic resonance Fano resonance was described in a general form. Therefore, it is applicable to plasmonic resonators. Here we see two examples.

Figure 2.6 shows an example of plasmon-related Fano resonance (Giannini et al., 2011). Figure 2.6a show top-view SEM images of fabricated Au structure of nanorod-antenna array: a wide view is provided in the upper panel with scale bar of 1 μm and a magnified view is shown in the lower panel. Nanogaps less than 60 nm were prepared whereas the length of Au rod, L, was varied from 1 to 2 μm. As a result, the broad plasmonic resonances in the Au-rod array were excited at 1400–2400 cm^{-1} with linear polarization parallel to the long axis of the rod.

Figure 2.6b shows a polymer PMMA[c] film was made on the Au-rod array by spin coating. The C=O vibration mode in PMMA

[c] Polymethylmetacrylate.

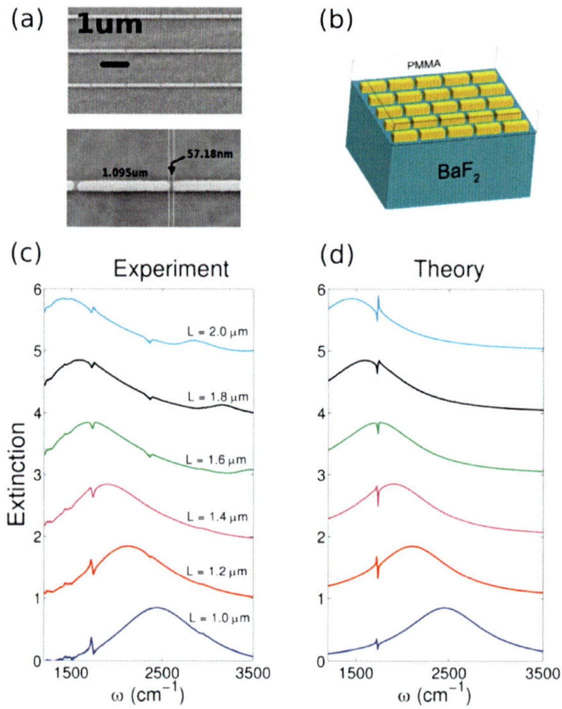

Figure 2.6 Fano resonance of a discrete molecular vibration mode at 1730 cm^{-1} and broad plasmonic resonance. (a) Top-view SEM images of specimen. (b) Illustration of a wide view of the array of plasmonic nanoantenna. (c) and (d) Extinction spectra in experiment and theory, respectively. Reprinted with permission from Giannini et al. (2011). Copyright 2011 American Chemical Society.

has large oscillator strength and appears 1730 cm^{-1}. Thus, a configuration where a discrete vibration mode interacts with a broad plasmonic resonance was prepared.

Figure 2.6c shows measured extinction spectra of the system in Fig. 2.6b. The corresponding spectra evaluated by Fano-resonance theory were calculated and presented in Fig. 2.6d. The theoretical extinction spectra at the 1730 cm^{-1} qualitatively agreed with those in the measurement. The shapes depend on the energies of the discrete vibration mode and the broad plasmonic resonance. In general, the relative shapes of the narrow and broad resonances change depending on the coupling strengths of excitation light with the discrete state and with broad state (or continuum).

Analysis for optical spectra Fano resonance is also observed in optical spectra such as transmittance (T) and R. Note that extinction is proportional to absorption coefficient and therefore Fig. 2.6 is directly comparable to transition probability. We here address an example to analyze optical spectra on Fano-type resonances that need a modified analysis.

Complex transmittance t and complex reflectance r are defined by

$$t = \frac{E_{\mathrm{t}}}{E_{\mathrm{in}}} \tag{2.48}$$

$$r = \frac{E_{\mathrm{r}}}{E_{\mathrm{in}}} \tag{2.49}$$

using E-field components of incidence (E_{in}), transmission (E_t), and reflection (E_r).[d] They are generally complex-valued, including the information on phase of the EM waves.

Taking account of response from Fano-type resonances, the t and r are phenomenologically expressed as

$$t(\omega) = a_t + \sum_k \frac{f_k \Gamma_k}{\hbar\omega - E_k + i\Gamma_k} \tag{2.50}$$

$$r(\omega) = a_r \pm \sum_k \frac{f_k \Gamma_k}{\hbar\omega - E_k + i\Gamma_k} \tag{2.51}$$

where a_t and a_r are fitting parameters in the analysis, representing base line of the analyzing range. Also, f_k, E_k, and Γ_k are treated as parameters in fitting though they have physical meanings as oscillator strength, resonant energy, and line width, respectively. The phenomenological Lorentzian model for spectrum shape was reported (Fan and Joannopoulos, 2002) and applied to analyze the optical responses of plasmon hybrid systems though they were different from the original configuration composed of a narrow discrete state and a broad (or, ideally, continuous) states in Fig. 2.3. Fano-shape optical responses in the T spectra (Ropers et al., 2005) and R spectra (Christ et al., 2007, 2008) were analyzed.

---

[d] Here we mainly think of subwavelength structures free from diffraction.

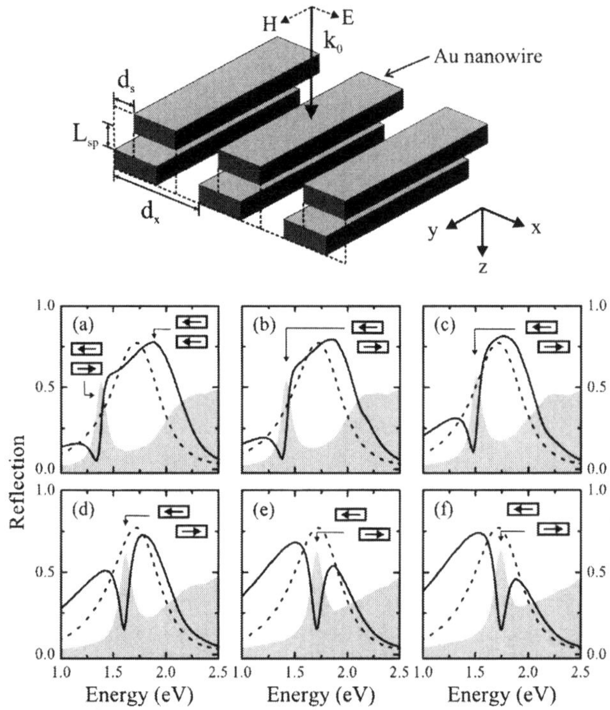

Figure 2.7 Analysis for optical spectra using Fano-resonance model. R($=|r|^2$) spectra were analyzed using Eq. 2.51. Stacked Au-nanorod array is illustrated at the top. (a)–(f) Numerically calculated R (solid curves) and absorbance (gray) spectra in various configurations changing lateral displacement d_s of the stacked nanorod pair. The absorbance spectrum of the single-layer nanorod array is shown with dashed curves. Reprinted with permission from Christ et al. (2007). Copyright 2007 by the American Physical Society.

Figure 2.7 shows an example of R-spectra analysis using Fano-resonance model (Christ et al., 2007). Note that R in Fig. 2.7 is displayed in the 0–1 range and R $= |r|^2$. Each Au nanorod was set to have section of 100×15 nm^2 and was periodically arrayed at a periodicity d_x of 200 nm along the x axis. The nanorods were assumed to be infinitely long along the y axis. In the numerical calculation based on the RCWA+S-matrix method, the Au nanorods were set to be embedded in SiO_2 matrix.

Figures 2.7a–f present a series of R spectra (black solid curves) with setting the displacement d_s to 0, 20, 40, 60, 80, and 100 nm, respectively. The distance of the two periodic layers, L_{sp}, was fixed at 30 nm. Incident light travels along the z axis and was set to x polarization. Absorbance spectra are also shown with gray shadow. At $d_s = 0$ nm, an asymmetric R dip at 1.3 eV and a R peak at 1.75 eV correspond to plasmonic resonances of out-of-phase and in-phase oscillation, as indicated in Fig. 2.7a. The two resonances appear at the lower and higher energies than the resonance of single-layer Au nanorod array (dashed curve). This behavior is ascribed to plasmon–plasmon coupling and understood from analogy of bonding and anti-bonding states of a two-atom molecule.

In terms of Fano resonance, the original two resonances are same to each other, so that the parameter q takes $q \approx 1$ from the definition in Eq. 2.43. The parameter q suggests asymmetric response around the original resonant energy and indeed the R spectra in Figs. 2.7a–f are deeply asymmetric around the resonant dips. As the displacement d_s increased, the resonant energy of the out-of-phase oscillation shifts to higher energies; finally, at $d_s = 100$ nm, the resonant energy reach 1.75 eV. The R spectrum was reproduced by Eq. 2.51 (Christ et al., 2007), in which the sign at the right-hand side of Eq. 2.51 was set to $-$. As was shown in this example, Fano-type resonant model is valid even for the coupling systems of identical resonance. In other words, the Fano resonance takes place, irrespective of the line widths of the original resonances. In this case, plasmon–plasmon coupling (or hybridization of plasmons) takes place. Thus, the Fano resonance is a quite general description for resonances observed in materials and artificial structures.

We also mention strong-coupling of the plasmonic resonances resident in the stacking two nanorods. Figure 2.7a indicates a large energy splitting of anti-parallel and parallel oscillations in the Au-nanorod pair. The oscillation modes exhibit anti-crossing dispersion in $(k_{\|}, \omega)$ plane where $k_{\|}$ is wavenumber parallel to the xy plane. The energy splitting and anti-crossing dispersion support the strong coupling of the resonances in the two nanorods. As the displacement d_s increases, the lower resonant energy of the coupled mode shifts to the energy of single-layer nanorod array at 1.75 eV, shown with the dashed curve; this tendency implies that near-field coupling between the stacking nanorods contributes to the strong coupling.

Strong-coupling systems of plasmon with another resonance are addressed later (see Sections 5.1 and 5.5). Concrete examples of plasmon–plasmon coupling are shown in Chapter 3: metallic dimer hybridization (Fig. 3.16) and heteroplasmon coupling (Fig. 3.46).

2.4 Generalized Theory for Response Function

MMs studies greatly stimulated theoretical interest in the framework on macroscopic material parameters such as permittivity ε and permeability μ. In this section, we briefly overview the consequences of response function based on microscopic Maxwell equations that are logically consistent with the quantum electromagnetism.

Basic framework Full quantum mechanical treatment is not available, at present, for systems of condensed matter and EM waves, because solids and liquids are not solved in the first-principles manner. Therefore, semiclassical, microscopic, and nonlocal theory was constructed in a practical point of view (Cho, 2003, 2010).

The main framework of the semiclassical theory is as follows.

(i) EM fields are described in a classical way.
(ii) Matter is described in the non-relativistic quantum mechanics.
(iii) EM fields are formulated based on vector and scalar potentials, $\mathbf{A}$ and ϕ, respectively. When we take Coulomb gauge, the vector potential satisfies

$$\nabla \cdot \mathbf{A} = 0. \tag{2.52}$$

The $\mathbf{A}$ is decomposed into transverse $\mathbf{A}^{(T)}$ and longitudinal $\mathbf{A}^{(L)}$ components such as

$$\mathbf{A} = \mathbf{A}^{(T)} + \mathbf{A}^{(L)}.$$

Under the Coulomb gauge

$$\mathbf{A}^{(L)} = \mathbf{0}$$

and therefore the following relation such that

$$\mathbf{A} = \mathbf{A}^{(T)} \tag{2.53}$$

holds and means that the vector potential is purely composed of transverse component.

The Hamiltonian $\mathcal{H}$ in the semiclassical, microscopic, and nonlocal theory is expressed as

$$\mathcal{H} = \mathcal{H}_{\text{matEM}} + \mathcal{H}_{\text{EM}} + \mathcal{H}_{\text{int}}. \tag{2.54}$$

Each term in the right-hand side of Eq. 2.54 is explicitly written in the following, using Coulomb energy $\mathcal{U}_{\text{Coulomb}}$ and current density $\mathbf{J}_0$.

$$\mathcal{H}_{\text{matEM}} = \sum_l \frac{\{\mathbf{p}_l - e_l \mathbf{A}(\mathbf{r}_l)/c_0\}^2}{2m_l} + \mathcal{U}_{\text{Coulomb}} \tag{2.55}$$

$$\mathcal{U}_{\text{Coulomb}} = \frac{1}{4\pi\varepsilon_0} \sum_l \sum_{k>l} \frac{e_k e_l}{|\mathbf{k}_l - \mathbf{r}_l|} \tag{2.56}$$

$$\mathcal{H}_{\text{EM}} = \frac{\varepsilon_0}{2} \int \left[\left\{ \frac{\partial \mathbf{A}(\mathbf{r})}{\partial t} \right\}^2 + c_0^2 \{\nabla \times \mathbf{A}(\mathbf{r})\}^2 \right] d\mathbf{r} \tag{2.57}$$

$$\mathcal{H}_{\text{int}} = -\int \mathbf{J}_0(\mathbf{r}) \cdot \mathbf{A}(\mathbf{r}) d\mathbf{r} \tag{2.58}$$

$$\mathbf{J}_0(\mathbf{r}) = \sum_l \frac{e_l}{2m_l} \{\mathbf{p}_l \delta(\mathbf{r} - \mathbf{r}_l) + \delta(\mathbf{r} - \mathbf{r}_l)\mathbf{p}_l\} \tag{2.59}$$

Note that the interaction term in Eq. 2.58 is able to include higher order interaction just by adding the term to the right-hand side of Eq. 2.58. For example, quadratic term is written as

$$\frac{1}{2} \int \hat{N}(\mathbf{r}) A(\mathbf{r})^2 d\mathbf{r} \tag{2.60}$$

where $\hat{N}$ is defined such that

$$\hat{N}(\mathbf{r}) = \sum_l \frac{e_l^2}{m_l} \delta(\mathbf{r} - \mathbf{r}_l). \tag{2.61}$$

Other interaction terms originating from orbital and spin also need to be included. The orbital current density operator $\mathbf{J}_{\text{orb}}$ is expressed as

$$\mathbf{J}_{\text{orb}} = \mathbf{J}_0 - \hat{N}(\mathbf{r}) \mathbf{A}(\mathbf{r}). \tag{2.62}$$

Spin-induced current density $\mathbf{J}_{\text{s}}$ is given by

$$\mathbf{J}_{\text{s}} = \nabla \times \mathbf{M}_{\text{s}}(\mathbf{r}) \tag{2.63}$$

where spin magnetic polarization $\mathbf{M}_s$ has explicit form such that

$$\mathbf{M}_s(\mathbf{r}) = \sum_l \frac{e}{m_0 c_0} \hbar\vec{\sigma}_l \delta(\mathbf{r} - \mathbf{r}_l).$$

The $\hbar\vec{\sigma}_l$ is spin angular momentum of an electron. Total current density $\mathbf{I}_{\mathrm{tot}}$ is written as

$$\mathbf{I}_{\mathrm{tot}} = \mathbf{J}_{\mathrm{orb}} + \mathbf{J}_{\mathrm{s}} \tag{2.64}$$

$$= \mathbf{I}(\mathbf{r}) - \hat{N}(\mathbf{r})\mathbf{A}(\mathbf{r}). \tag{2.65}$$

Equation 2.65 explicitly decomposes **A**-independent and **A**-dependent terms, and defines general current density including the spin-induced current density.

From Eqs. 2.62–2.65, we reach the expression on the $\mathbf{I}(\mathbf{r})$ such that

$$\mathbf{I}(\mathbf{r}) = \mathbf{J}_0(\mathbf{r}) + \mathbf{J}_s(\mathbf{r}). \tag{2.66}$$

Equation 2.66 is used to derive susceptibility in the present microscopic nonlocal framework.

We here provide general remarks regarding the theory in this section.

(i) Although relativistic correction for matter Hamiltonian is not addressed, the correction (i.e., spin-orbit, mass–velocity, and Darwin terms) can be incorporated, if necessary, into the matter Hamiltonian $\mathcal{H}_{\mathrm{mat}}$ such that

$$\mathcal{H}_{\mathrm{mat}} = \sum_l \frac{\mathbf{p}_l^2}{2m_l} + \mathcal{U}_{\mathrm{Coulomb}}, \tag{2.67}$$

which is obtained by setting $\mathbf{A}(\mathbf{r}_l) = \mathbf{0}$ in Eq. 2.55 (Cho, 2010).

(ii) We are mainly considering plasmonic resonances at optical frequencies. Therefore, spin-related terms in the Hamiltonian $\mathcal{H}$ do not show any prominent effect. Still, the theory without loss of generality is important as a basis of physical understanding and therefore we are referring to the theory in this section.

(iii) The expressions in Eqs. 2.55–2.59 look as if electrons (or charged particles) are involved as quasi-free particles at a glance. However, electrons in solids (or medium) are, in general, not quasi-free particles; strictly, the electrons in Eqs. 2.55–2.59 should be replaced with electronic eigenstates in solids.

Derivation of linear susceptibility In the microscopic nonlocal description, let us derive the susceptibility.

Linear matter–EM-field interaction is generally written as

$$\mathcal{H}_{\rm int} = -\int \mathbf{I}(\mathbf{r}) \cdot \mathbf{A}(\mathbf{r}) d\mathbf{r}. \tag{2.68}$$

Note that Eq. 2.68 manifests local interaction of matter and transverse component of EM field. In the derivation for the susceptibility, time-dependent perturbation is conducted. Schrödinger equation is given by

$$i\hbar \frac{\partial \Psi(t)}{\partial t} = (\mathcal{H}_0 + \mathcal{H}_{\rm int})\Psi(t) \tag{2.69}$$

where $\mathcal{H}_0$ denotes non-interacting Hamiltonian, equal to the sum of $\mathcal{H}_{\rm mat}$ Eq. 2.67 and the relativistic correction terms $\mathcal{H}_{\rm rc}$, that is,

$$\mathcal{H}_0 = \mathcal{H}_{\rm mat} + \mathcal{H}_{\rm rc}. \tag{2.70}$$

The interaction representation $\Psi_I(t) = \exp(i\mathcal{H}_0 t/\hbar)\Psi(t)$ enables us to transform Eq. 2.69 into

$$i\hbar \frac{\partial \Psi_I(t)}{\partial t} = \mathcal{H}_{\rm int}(t)\Psi_I(t). \tag{2.71}$$

Here we define time-dependent $\mathcal{H}_{\rm int}(t)$ as

$$\mathcal{H}_{\rm int}(t) = \exp(i\mathcal{H}_0 t/\hbar)\mathcal{H}_{\rm int} \exp(-i\mathcal{H}_0 t/\hbar). \tag{2.72}$$

Let us expand the wave function $\Psi(t)$ such that

$$\Psi(t) = \sum_\nu a_\nu(t)|\nu\rangle \tag{2.73}$$

where $|\nu\rangle$ represents eigenstate of matter Hamiltonian $\mathcal{H}_0(= \mathcal{H}_{\rm mat}+$ relativistic correction), satisfying

$$\mathcal{H}_0|\nu\rangle = E_\nu|\nu\rangle.$$

The lowest-order coefficient has the following solution such that

$$a_0(t) = \exp(-i\omega_0 t) \tag{2.74}$$

where $\omega_0 = E_0/\hbar$. The other coefficient $a_\nu(t)$ $(\nu \neq 0)$ is expressed as

$$a_\nu(t) = \frac{i}{\hbar}\int_{-\infty}^{t} ds\langle\nu|\exp(-i\mathcal{H}_0 t/\hbar)\mathcal{H}_{\rm int}(s)\exp(\gamma s/2)|0\rangle \tag{2.75}$$

where γ is adiabatic parameter of $\gamma > 0$. When we use Fourier-expanded formula for $\mathbf{A}(\mathbf{r}, t)$ such that

$$\mathbf{A}(\mathbf{r}, t) = \sum_{\omega} \mathbf{A}(\mathbf{r}, \omega) \exp(-i\omega t), \tag{2.76}$$

Eq. 2.76 is transformed into

$$a_\nu(t) = \frac{-1}{\hbar} \sum_{\omega} \frac{\exp\{-i(\omega_0 + \omega + i\gamma/2)t\}}{\omega_{\nu 0} - \omega - i\gamma/2} \int \langle \nu | \mathbf{I}(\mathbf{r}) | 0 \rangle \cdot \mathbf{A}(\mathbf{r}, \omega) d\mathbf{r} \tag{2.77}$$

where $\omega_{\nu 0} = E_\nu - E_0$.

Collecting all the **A**-linear terms in induced current density $\langle \Psi(t) | \mathbf{I}_{\text{tot}} | \Psi(t) \rangle$, one can define nonlocal susceptibility χ in the following equation:

$$\mathbf{I}_{\text{tot}}(\mathbf{r}, \omega) = \int d\mathbf{r}' \chi(\mathbf{r}, \mathbf{r}', \omega) \cdot \mathbf{A}(\mathbf{r}', \omega). \tag{2.78}$$

Note that Eq. 2.78 is a constitutive equation for the basic variables **A** and $\mathbf{I}_{\text{tot}}$, which corresponds to the conventional constitutive equations for **D**, **E**, **B**, and **H** (Eqs. 1.11 and 1.12), and provides a more unified representation. Finally, the susceptibility χ has the explicit form as follows.

$$\begin{aligned} \chi(\mathbf{r}, \mathbf{r}', \omega) &= -\langle 0 | \hat{N}(\mathbf{r}) | 0 \rangle \delta(\mathbf{r} - \mathbf{r}') \\ &+ \sum_{\nu} \left\{ \frac{\hbar^{-1} \mathbf{I}_{0\nu}(\mathbf{r}) \mathbf{I}_{\nu 0}(\mathbf{r}')}{\omega_{\nu 0} - \omega - i\gamma/2} + \frac{\hbar^{-1} \mathbf{I}_{\nu 0}(\mathbf{r}) \mathbf{I}_{0\nu}(\mathbf{r}')}{\omega_{\nu 0} + \omega + i\gamma/2} \right\} \end{aligned} \tag{2.79}$$

where matrix element of current operator is defined as

$$\mathbf{I}_{fi}(\mathbf{r}) = \langle f | \mathbf{I}(\mathbf{r}) | i \rangle. \tag{2.80}$$

One of the advantages in the expression of $\chi(\mathbf{r}, \mathbf{r}', \omega)$ is that the second term in the right-hand side of Eq. 2.79 comprises the product of **r**- and **r**′-dependent terms. The separated form makes it substantially feasible to solve the integral equation including the $\chi(\mathbf{r}, \mathbf{r}', \omega)$.

The second term of the right-hand side in Eq. 2.79 is approximated around $\omega \approx \omega_{\nu 0}$ such as

$$\frac{F}{\omega_{\nu 0}^2 - (\omega + i\gamma/2)^2} \approx \frac{F}{\omega_{\nu 0}^2 - \omega^2 - i\gamma\omega} \tag{2.81}$$

under the condition of $|\gamma/\omega| \ll 1$. In Eq. 2.81, factors except for the denominators were written as F. Note that Eq. 2.81 indicates Lorentz-type response function, equivalent to Eq. 2.17, and supports the classical Lorentz model in Section 2.1.

Generalized dispersion equation We briefly derive dispersion equation in the present generalized form. The vector potential **A** obeys wave equation, which is Maxwell wave equation in the **A** representation, such that

$$-\Delta\mathbf{A}(\mathbf{r}, t) + \frac{1}{c_0^2}\frac{\partial^2\mathbf{A}(\mathbf{r}, t)}{\partial t^2} = \mu_0\mathbf{I}_{\text{tot}}^{(T)}(\mathbf{r}, t) \tag{2.82}$$

where $\mathbf{I}_{\text{tot}}^{(T)}$ is the transverse component of $\mathbf{I}_{\text{tot}}$. We remind that the vector potential **A** is purely transverse because of the Coulomb gauge Eq. 2.53. Equation 2.82 is Fourier-transformed as follows.

$$(k^2 - k_0^2)\mathbf{A}(\mathbf{k}, \omega) = \mu_0\mathbf{I}_{\text{tot}}^{(T)}(\mathbf{k}, \omega) \tag{2.83}$$

where $k = |\mathbf{k}|$ and $k_0 = \omega/c_0$.

Let us recall another basic equation, Eq. 2.78, which is the generalized constitutive equation containing both electric and magnetic constitutive equations. Equation 2.78 is expressed, after Fourier transformation, as

$$\mathbf{I}_{\text{tot}}^{(T)}(\mathbf{k}, \omega) = \chi^{(T)}(\mathbf{k}, \omega)\mathbf{A}(\mathbf{k}, \omega). \tag{2.84}$$

The transverse component $\chi^{(T)}$ is defined by

$$\chi^{(T)}(\mathbf{k}, \omega) = (\mathbf{1} - \hat{\mathbf{k}}\hat{\mathbf{k}}) \cdot \chi(\mathbf{k}, \omega) \tag{2.85}$$

where $\hat{\mathbf{k}} = \mathbf{k}/|\mathbf{k}|$. Combining Eqs. 2.83 and 2.84, we reach the dispersion equation such that

$$\det\left[\frac{k^2}{k_0^2}\mathbf{1} - \left\{\mathbf{1} + \frac{1}{\varepsilon_0\omega^2}\chi^{(T)}(\mathbf{k}, \omega)\right\}\right] = 0. \tag{2.86}$$

The matrices in Eq. 2.86 are 2×2 dimensions.

Equation 2.86 contains all the transverse EM modes while it does not include the longitudinal modes. We note the followings for Eq. 2.86.

(i) All the poles are single in the complex frequency plane. This is obvious from the concrete expression of χ (Eq. 2.79).

(ii) The derivation of Eq. 2.86 was assumed to be subdiffraction condition because of simplicity. When diffraction channels are open, Eq. 2.86 becomes bigger than the 2×2 matrix. This subdiffraction assumption is a little more general than the long-wavelength approximation (Cho, 2010).

Description for longitudinal components As noted in Section 2.2, the longitudinal excitation can be induced at arbitrary frequencies ω and wavenumbers k. In this section, because of the Coulomb gauge (Eq. 2.52), the vector potential for EM waves has been assumed to be purely transverse (Eq. 2.53). Thus, to include the longitudinal component, additional treatment is necessary as follows.[e]

Interaction energy for the longitudinal component is written, using longitudinal polarization $\mathbf{P}^{(L)}$ (defined later) with longitudinal E-field vector $\mathbf{E}^{(L)}$, such that

$$-\int \mathbf{P}^{(L)} \cdot \mathbf{E}^{(L)} d\mathbf{r} = \iint d\mathbf{r}d\mathbf{r}' \frac{\rho(\mathbf{r})\rho(\mathbf{r}')}{|\mathbf{r}-\mathbf{r}'|} \tag{2.87}$$

where ρ is charge density, defined by

$$\rho(\mathbf{r}) = \sum_l e_l \delta(\mathbf{r}-\mathbf{r}'). \tag{2.88}$$

The longitudinal polarization $\mathbf{P}^{(L)}$ is produced from the following equation such that

$$\nabla \cdot \mathbf{P}^{(L)}(\mathbf{r}) = -\rho(\mathbf{r}). \tag{2.89}$$

The charge density ρ generally contains both genuine and induced currents. The solution of Poisson-type equation is obtained as follows.

$$\mathbf{P}^{(L)}(\mathbf{r}) = \nabla \int d\mathbf{r}' \frac{\rho(\mathbf{r}')}{4\pi|\mathbf{r}-\mathbf{r}'|} \tag{2.90}$$

On the other hand, $\mathbf{E}^{(L)}$ is the sum of incident E field $\mathbf{E}^{(L)}_{\rm ext}$ and the charge-induced longitudinal fields, and is written as

$$\mathbf{E}^{(L)}(\mathbf{r}) = \mathbf{E}^{(L)}_{\rm ext}(\mathbf{r}) - \nabla \int d\mathbf{r}' \frac{\rho(\mathbf{r}')}{4\pi\varepsilon_0|\mathbf{r}-\mathbf{r}'|}. \tag{2.91}$$

Equations 2.90 and 2.91 suggest that the longitudinal components are able to take finite values even when the charge density ρ consists only of the induced part; this situation usually takes place in plasmonic resonators.

---

[e]This additional treatment is inevitable under the Coulomb gauge. If we take the Lorentz gauge instead of the Coulomb gauge, this treatment is unnecessary. However, the susceptibility χ in Eq. 2.79 is no longer separable form in the Lorentz gauge, so that the integral equation including χ generally becomes difficult to be solved.

The interaction Hamiltonian $\mathcal{H}_{\text{int}}$ needs to be replaced with the corrected interaction Hamiltonian $\tilde{\mathcal{H}}_{\text{int}}$ such that

$$\tilde{\mathcal{H}}_{\text{int}} = \mathcal{H}_{\text{int}} - \int \mathbf{P}^{(L)} \cdot \mathbf{E}^{(L)} d\mathbf{r}. \tag{2.92}$$

Accordingly, the non-interacting Hamiltonian $\mathcal{H}_0$ (Eq. 2.70) is needed to be replaced with the sum of $\mathcal{H}_{\text{matEM}}$ (Eq. 2.55), Coulomb energy of the longitudinal components $\mathcal{H}_{\text{cl}}$, and relativistic corrections $\mathcal{H}_{\text{rc}}$; the corrected non-interacting Hamiltonian $\tilde{\mathcal{H}}_0$ is written as

$$\tilde{\mathcal{H}}_0 = \mathcal{H}_{\text{matEM}} + \mathcal{H}_{\text{cl}} + \mathcal{H}_{\text{rc}}. \tag{2.93}$$

The corrections of the Hamiltonians result in corrected constitutive equation containing both transverse and longitudinal components:

$$\mathbf{I}_{\text{tot}}(\mathbf{r}, \omega) = \int d\mathbf{r}' \left\{ \chi(\mathbf{r}, \mathbf{r}', \omega) \cdot \mathbf{A}(\mathbf{r}', \omega) + \chi_L(\mathbf{r}, \mathbf{r}', \omega) \mathbf{E}^{(L)}(\mathbf{r}', \omega) \right\}. \tag{2.94}$$

The longitudinal susceptibility χ_L is thus introduced.

Note that the field variable is not $\mathbf{A}$ but $\mathbf{E}^{(L)}$ for the longitudinal component. Also, Eq. 2.91 for the $\mathbf{E}^{(L)}$ and Eq. 2.94 for the constitutive equation cannot be combined such as the dispersion equation for the transverse component (Eq. 2.86); therefore, there is not an explicit constraint for the longitudinal component in the (k, ω) plane, as mentioned at the end of Section 2.2.

The inclusion of the longitudinal component is possible in the Coulomb gauge though some effort is needed and the conciseness is lost as shown above. The longitudinal component is mostly neglected in textbooks; however, it is not negligible in plasmonic resonators.

Derivation of nonlinear susceptibility One of the advantages in the microscopic nonlocal theory is that nonlinear responses are naturally derived through the perturbation procedure. Let us mention nonlinear susceptibility.

Following a standard perturbation using density matrix, the higher-order (i.e., nonlinear) expectation values of arbitrary operators are determined in an iterative manner. As we define matrix-from factor $F_{\mu\nu}$ by

$$F_{\mu\nu} = \int d\mathbf{r} \langle \mu | \mathbf{I}(\mathbf{r}) | \nu \rangle \cdot \mathbf{A}(\mathbf{r}, \omega), \tag{2.95}$$

the third-order nonlinear current density $\mathbf{J}^{(3)}$ is obtained through a perturbation procedure and finally expressed as

$$\mathbf{J}^{(3)}(3\omega) = \frac{-1}{\hbar^3} \sum_{\mu,\nu,\sigma} \frac{F_{0\mu}(\omega)F_{\mu\nu}(\omega)F_{\nu\sigma}(\omega)\langle\sigma|\mathbf{I}(\mathbf{r})|0\rangle}{(\omega_{0\mu}-\omega)(\omega_{0\nu}-2\omega)(\omega_{0\sigma}-3\omega)}$$
$$+(\mu\nu\sigma\text{-permutated other 7 terms}). \qquad (2.96)$$

The third-order nonlinear susceptibility $\chi^{(3)}$ is determined from Eq. 2.96. Note that a more general case of $3\omega \rightarrow \omega_1 + \omega_2 + \omega_3$ is expressed in a similar form to Eq. 2.96.

Experimental evidence to support the nonlinear response was reported regarding optical four-wave mixing signals from multi-quantum well of semiconductors at an exciton-induced range (Ishihara et al., 2002). It was shown that the size in nm dimensions contributes to resonant energy.

The general response function theory in this section is a rational framework. Still, the actual examples to analyze experimental data are not so many, so that it does not seem that the theoretical method has been widely employed. It is therefore preferred that the theory will be established through successful analyses for various linear and nonlinear cases.

2.5 Summary

In this chapter, we addressed the theory of response function χ, which is fundamental physics on resonances in materials and artificial structures including plasmonic resonators.

In Sections 2.1 and 2.2, dielectric function $\varepsilon(\omega)$, which is defined using the response function χ, was derived in the classical and quantum mechanical manners, respectively. Both regimes reached the same dielectric function (Eqs. 2.14 and 2.29). The fundamental properties and the relation to longitudinal excitations were described.

In Section 2.3, Fano resonance was described to reproduce experimental data. The Fano-type spectral shapes on resonances was shown to be a direct consequence of configuration interaction of a discrete state with continuum. The illustrative examples of the analyses were specified (Figs. 2.5 and 2.6). Furthermore, optical spectra coming from plasmon–plasmon couplings were described as Fano-type resonances (Fig. 2.7). It was thus shown that Fano resonances can be applied to quite general configurations that two resonances become coupled.

In Section 2.4, generalized response function in a contemporary fashion was addressed. The microscopic nonlocal formalism was recently proposed in a framework to keep the logical consistency in the physics of light–matter interaction. Susceptibility was derived for linear case (Eq. 2.79). The third-order nonlinear current density was noted (Eq. 2.96). In addition, the transverse and longitudinal components in the linear susceptibility were described (Eq. 2.94); it is important that both components contribute to plasmonic resonances. The dispersion equation was explicitly shown for the transverse component (Eq. 2.86) whereas it was pointed out that the explicit dispersion equation does not hold for the longitudinal component.

References

Cho, K. (2003). *Optical Response of Nanostructures: Microscopic Nonlocal Theory* (Springer, Berlin).

Cho, K. (2010). *Reconstruction of Macroscopic Maxwell Equations* (Springer, Berlin).

Christ, A., Ekinci, Y., Solak, H. H., Gippius, N. A., Tikhodeev, S. G., and Martin, O. J. F. (2007). Controlling the fano interference in a plasmonic lattice, *Phys. Rev. B* **76**, 201405(R).

Christ, A., Martin, O. J. F., Ekinci, Y., Gippius, N. A., and Tikhodeev, S. G. (2008). Symmetry breaking in a plasmonic metamaterial at optical wavelength, *Nano. Lett.* **8**, 2171–2175.

Dirac, P. (1927). Über die quantenmechanik der stoßvorgänge, *Z. Physik* **44**, 585–595.

Fan, S., and Joannopoulos, J. D. (2002). Analysis of guided resonances in photonic crystal slabs, *Phys. Rev. B* **65**, 235112.

Fano, U. (1956). Atomic theory of electromagnetic interactions in dense materials, *Phys. Rev.* **103**, 1202–1218.

Fano, U. (1961). Effects of configuration interaction on intensities and phase shifts, *Phys. Rev.* **124**, 1866–1878.

Giannini, V., Francescato, Y., Amrania, H., Phillips, C. C., and Maier, S. A. (2011). Fano resonances in nanoscale plasmonic systems: A parameter-free modeling approach, *Nano Lett.* **11**, 2835–2840.

Hopfield, J. J. (1958). Theory of the contribution of excitons to the complex dielectric constant of crystals, *Phys. Rev.* **112**, 1555–1567.

Ishihara, H., Cho, K., Akiyama, K., Tomita, N., Nomura, Y., and Isu, T. (2002). Large four-wave mixing of spatially extended excitonic states in thin GaAs layers, *Phys. Rev. Lett.* **89**, 017402.

Jackson, J. D. (1999). *Classical Electrodynamics*, 3rd edn. (Wiley, Hoboken, NJ).

Kobayashi, K. (1997). *Hikari Bussei Numon (Introduction to Optical Properties in Solids)* (Shokabo, Tokyo), in Japanese.

Landau, L. D., Lifshitz, E. M., and Pitaevskii, P. (1984). *Electrodynamics of Continuous Media*, 2nd edn. (Pergamon, New York).

Luk'yanchuk, B., Zheludev, N. I., Maier, S. A., Halas, N. J., Nordlander, P., Giessen, H., and Chong, C. T. (2010). The Fano resonance in plasmonic nanostructures and metamaterials, *Nature Mater.* **9**, 707–715.

Miroshnichenko, A. E., Flach, S., and Kivshar, Y. S. (2010). Fano resonances in nanoscale structures, *Rev. Mod. Phys.* **82**, 2257–2298.

Ropers, C., Park, D. J., Stibenz, G., Steinmeyer, G., Kim, J., Kim, D. S., and Lienau, C. (2005). Femtosecond light transmission and subradiant damping in plasmonic crystals, *Phys. Rev. Lett.* **94**, 113901.

Wooten, F. (1972). *Optical Properties of Solids* (Academic Press, New York).

Chapter 3

Plasmonic Resonators

Numerous studies have yielded much variety in plasmonic resonators. Figure 3.1 shows classification list of plasmonic resonators in terms of the key elements and layer structures, indicating the corresponding sections. Readers can easily confirm that all the

Key elements \ Layer structures	Single	Simply stacked	Stacked complementary (SC)
NPs Wires SRRs	Sec. 3.2, 3.3 3.5, 3.7 3.4.2 (1.4.3)	Sec. 3.4.2, 3.9	Sec. 3.8
Holes	Sec. 3.4.1	Sec 3.6.4	Sec. 3.8
Metallic films	Sec. 3.1.1, 3.1.2 (1.4.1)	Sec. 3.1.3, 3.6, 3.9	×

Figure 3.1 Classification of plasmonic resonators from the key elements and layer structures. The corresponding sections are indicated. Symbol × indicates logically impossible structure. Possible combinations of the key elements and layer structures are fully covered in this chapter. NPs mean nanoparticles. Relevant sections in Chapter 1 are included.

Plasmonic Resonators: Fundamentals, Advances, and Applications
Masanobu Iwanaga

ISBN 978-981-4745-06-2 (Hardcover), 978-981-4745-07-9 (eBook)
www.panstanford.com

logically conceivable plasmonic resonators are contained, which is a feature in this book.

In this chapter, they are classified into the type of resonances and the features are described.

In Sections 3.1–3.3, we address individual plasmonic resonators such as plasmonic waveguides, metallic nanoparticles (NPs), and the assembles. These have been widely recognized as plasmonic structures and their hot spots were often emphasized as the feature.

In Sections 3.4–3.8, we describe periodic structures of plasmonic resonators such as single-layer lattices of perforated metallic film, simply stacked structures, and stacked complementary (SC) structures. For incident plane waves, these structures generally exhibit reflection, transmission, and diffraction; when the periodicity is adequately smaller than the incident wavelength, they respond to the incident light just as solid or liquid materials without diffraction, and the macroscopic optical responses characterize their resonant properties as described in Chapter 2.

In addition to the individual plasmonic resonators and the arrays, we also refer to perfect absorbers (PAs) in Section 3.9, which leads plasmonic resonators to various applications, and to plasmonic continuum in Section 3.5, which is usually observed in experiment but is hardly recognized consciously.

3.1 Plasmonic Waveguides

Let us start to survey plasmonic waveguides composed of a pair (or pairs) of flat metallic layers, which are often metallic thin films. The structures are often called as metal-insulator-metal (MIM) structures. The basics were clarified in 1960s (Swihart, 1961; Economou, 1969).

As described in Chapter 1, the study of surface plasmon polaritons was extensively conducted since 1970s; therefore, the MIM waveguides were investigated not in terms of plasmons but in terms of waveguides modes at the metal-insulator interface, and besides attracted interest in connection with superconductors that were successfully described by the BCS theory (Bardeen et al., 1957).

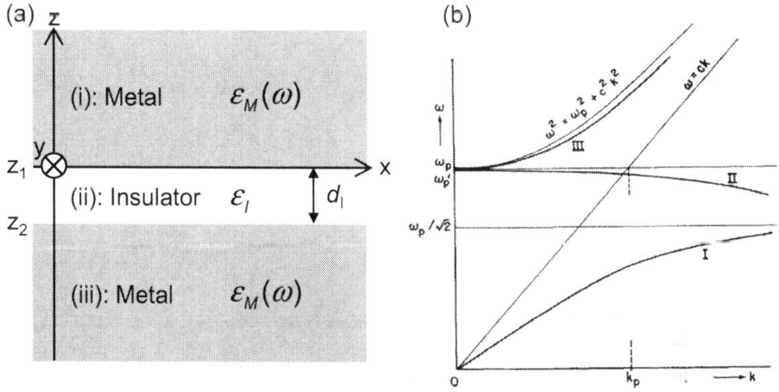

Figure 3.2 Single I layer between semi-infinitely thick M layers, that is, MIM structure. (a) Section-view schematic of the MIM structure and the coordinate. (b) Dispersion diagrams obtained from Eq. 3.11. (b) was reprinted with permission from Economou (1969). Copyright 1969 by the American Physical Society.

We have already addressed the surface plasmon polariton (SPP) at single MI interface in Section 1.3. Plasmonic waveguides of MIM, insulator-metal-insulator (IMI), and IMIMI structures are overviewed in this section.

3.1.1 *MIM: Single I Layer between Semi-Infinite M Layers*

Figure 3.2a shows a waveguide of single insulator (I) layer between semi-infinitely thick metal (M) layers. The three domains are labelled as (i), (ii), and (iii). Let us derive TM[a]-polarized solutions of MI-interface localized waves in this configuration. The solutions are generally expressed in each domain:

$$\text{(i)}\quad E_x(x,z) = E_{\text{i}}^{+}(x)\exp(\Gamma_M z) \tag{3.1}$$

$$\text{(ii)}\quad E_x(x,z) = E_{\text{ii}}^{+}(x)\exp(\Gamma_I z) + E_{\text{ii}}^{-}(x)\exp(-\Gamma_I z) \tag{3.2}$$

$$\text{(iii)}\quad E_x(x,z) = E_{\text{iii}}^{-}(x)\exp(-\Gamma_M z) \tag{3.3}$$

where $\text{Re}(\Gamma_M) < 0$ to ensure the relation of $|E_x(x,z)| \to 0$ as $z \to \pm\infty$, respectively. The sign $\pm$ denotes the propagation along the

[a]Transverse magnetic.

$\pm z$ directions. Note that Eqs. 3.1–3.3 are independent of y under the TM polarization.

The boundary condition for H_y component at the (i)–(ii) interface ($z = z_1$) is written as

$$\frac{i\varepsilon_0\varepsilon_M\omega}{\Gamma_M}E_{\mathrm{i}}^{+}(x)e^{\Gamma_M z_1} = \frac{i\varepsilon_0\varepsilon_I\omega}{\Gamma_I}\left(E_{\mathrm{ii}}^{+}(x)e^{\Gamma_I z_1} - E_{\mathrm{ii}}^{-}(x)e^{-\Gamma_I z_1}\right) \quad (3.4)$$

by using Eq. 1.49. Another boundary condition for E_x at $z = z_1$ is written as

$$E_{\mathrm{i}}^{+}(x)e^{\Gamma_M z_1} = E_{\mathrm{ii}}^{+}(x)e^{\Gamma_I z_1} + E_{\mathrm{ii}}^{-}(x)e^{-\Gamma_I z_1} \quad (3.5)$$

We divide Eq. 3.4 by Eq. 3.5 and obtain a relation such that

$$-R = \frac{E_{\mathrm{ii}}^{+}(x)e^{\Gamma_I z_1} - E_{\mathrm{ii}}^{-}(x)e^{-\Gamma_I z_1}}{E_{\mathrm{ii}}^{+}(x)e^{\Gamma_I z_1} + E_{\mathrm{ii}}^{-}(x)e^{-\Gamma_I z_1}} \quad (3.6)$$

where ratio R is defined by

$$R = -\frac{\varepsilon_M}{\Gamma_M}\frac{\Gamma_I}{\varepsilon_I}. \quad (3.7)$$

Similarly, the boundary condition at $z = z_2$ yields the following two equations.

$$\frac{i\varepsilon_0\varepsilon_M\omega}{-\Gamma_M}E_{\mathrm{iii}}^{-}(x)e^{-\Gamma_M z_2} = \frac{i\varepsilon_0\varepsilon_I\omega}{\Gamma_I}\left(E_{\mathrm{ii}}^{+}(x)e^{\Gamma_I z_2} - E_{\mathrm{ii}}^{-}(x)e^{-\Gamma_I z_2}\right) \quad (3.8)$$

$$E_{\mathrm{iii}}^{-}(x)e^{-\Gamma_M z_2} = E_{\mathrm{ii}}^{+}(x)e^{\Gamma_I z_2} + E_{\mathrm{ii}}^{-}(x)e^{-\Gamma_I z_2} \quad (3.9)$$

From Eq. 3.8 Eq. 3.9, we have

$$R = \frac{E_{\mathrm{ii}}^{+}(x)e^{\Gamma_I z_2} - E_{\mathrm{ii}}^{-}(x)e^{-\Gamma_I z_2}}{E_{\mathrm{ii}}^{+}(x)e^{\Gamma_I z_2} + E_{\mathrm{ii}}^{-}(x)e^{-\Gamma_I z_2}}. \quad (3.10)$$

By combining Eqs. 3.6 and 3.10 to eliminate $E_{\mathrm{ii}}^{\pm}$, we finally reach dispersion equation (DE) in this case:

$$\frac{1-R}{1+R} = \pm\exp(\Gamma_I d_I). \quad (3.11)$$

Figure 3.2b shows dispersion diagrams obtained from Eq. 3.11. One surface-wave solution, termed I, is an asymmetric waveguide mode in the MIM structure. Note that the gradient is smaller than $c(=c_0)$ even at the origin of the (k, ω) plane; besides, the gradient is dependent on the thickness d_I. This means that the mode I is not identical to the SPP at the single MI interface.

Other dispersion curves, termed II and III, start at $(0, \omega_p)$ where ω_p is the plasma frequency in Eq. 1.3. The solution II is a symmetric mode, appearing at a range of $(\omega_p/\sqrt{2}, \omega_p)$, and has negative group velocity. The solution III appears above ω_p and is an asymmetric mode though the dispersion is similar to a transverse wave of $\omega^2 = \omega_p^2 + c^2k^2$, which appeared as the upper branch in Fig. 1.9. As the thickness d_I of the insulator layer approaches to 0, the modes I and III disappear; still, the mode II remains and becomes close to $\omega = \omega_p$, which is bulk plasmon as one can easily imagine. These solutions were evaluated under the assumption for simplification such that $\varepsilon_I = 1$ and ε_M is Drude-metal permittivity in Eq. 1.24.

3.1.2 *IMI: Single M Layer between Semi-Infinite I Layers*

Figure 3.3a shows the configuration exchanging M and I in Fig. 3.2. The M slab of thickness d_M is also a waveguide though one might not recognized as a waveguide. The DE is obtained just by exchanging the suffix such as $M \leftrightarrow I$ in Section 3.1.1. Concretely, Eq. 3.7 is redefined by

$$R' = -\frac{\varepsilon_I}{\Gamma_I}\frac{\Gamma_M}{\varepsilon_M}. \tag{3.12}$$

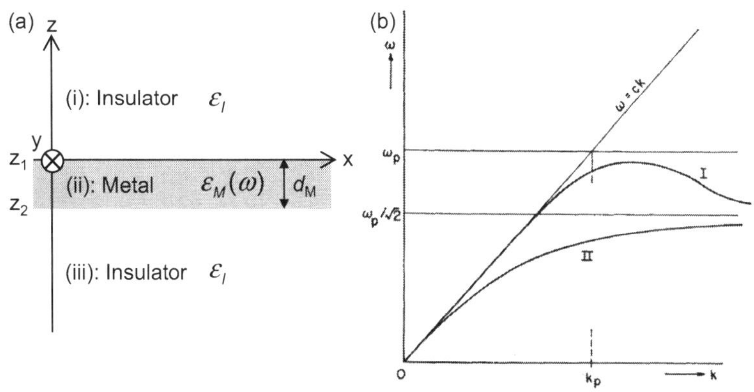

Figure 3.3 Single M layer between semi-infinitely thick I layers. (a) Schematic of IMI structure. (b) Dispersion diagrams obtained from Eq. 3.13. (b) was reprinted with permission from Economou (1969). Copyright 1969 by the American Physical Society.

The DE for the IMI structure is expressed as

$$\frac{1-R'}{1+R'} = \pm\exp(\Gamma_M d_M). \tag{3.13}$$

Note that, even if we use R instead of R' in Eqs. 3.12, and 3.13 is invariant.

Figure 3.13b shows dispersion diagrams of localized modes in the IMI structure. Two modes termed I and II were found to be the solutions (Economou, 1969); both of them are located below the light cone represented by $\omega = ck$, meaning that both are waveguide modes localized at the metallic slab. The mode II is a symmetric mode of SPP-like dispersion whereas the mode I is an asymmetric mode located at higher energies than the mode II. Note that, in this IMI configuration, there is no mode inside the light cone, in contrast to the MIM configuration in Fig. 3.2. In this case, too, the simplification was assumed such that $\varepsilon_I = 1$ and ε_M is Drude-metal permittivity in Eq. 1.24.

3.1.3 *IMIMI: MIM Layers between Semi-Infinite I Layers*

Figure 3.4 presents IMIMI structure, i.e., MIM structure between semi-infinitely thick I layers. The localized modes in the homogeneous MIM structure are exponentially decreasing solutions in the semi-infinitely thick insulator layers; the z dependence of E field is expressed as

$$E_\pm(z) = E_\pm \exp(\pm k_z^{(\pm)} z) \tag{3.14}$$

and $\mathrm{Re}(k_z^{(\pm)}) < 0$ where k_z is the z component of wavevector. The z dependence satisfies $E_\pm(z) \to 0$ as $z \to \pm\infty$, respectively. In the layers of finite thickness, the z dependence is written as

$$E_i(z) = E_i^+ \exp(ik_z^{(i)} z) + E_i^- \exp(-ik_z^{(i)} z) \tag{3.15}$$

where layer number $i = 1, 2, 3$. General expressions of Eqs. 3.14 and 3.15 are connected to each other at the boundaries of z_j ($j = 1, 2, 3, 4$). By eliminating the terms of $E_\pm$, E_i, $k_z^{(\pm)}$ and $k_z^{(i)}$, the DE for the localized modes is finally obtained and explicitly expressed

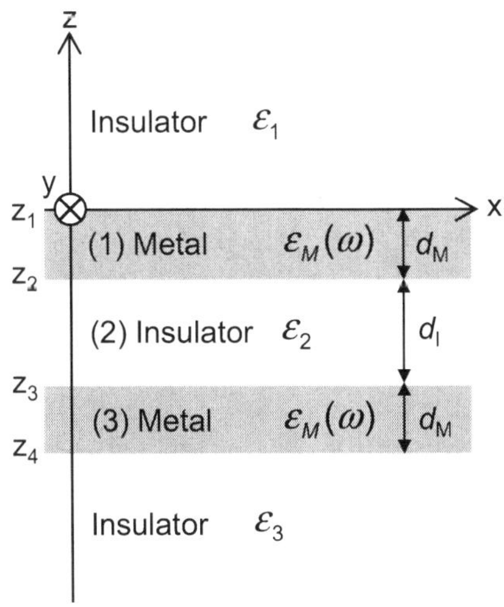

Figure 3.4 IMIMI structure. The three insulators have independent permittivity of ε_α ($\alpha = 1, 2, 3$). The permittivity and the thickness of the metal were set to be identical.

as follows:

$$\begin{aligned}
&\left\{\exp\left(2\sqrt{k_0^2\varepsilon_2 - k_\parallel^2}d_I\right) - 1\right\}\left(1 - \frac{Z_1^2}{Z_M^2}\right) \\
&\times\left(1 + \frac{Z_1 + Z_M}{Z_1 - Z_M}\frac{Z_M - Z_3}{Z_M + Z_3}\right) \\
&+\left\{\exp\left(2\sqrt{k_0^2\varepsilon_2 - k_\parallel^2}d_I\right)\left(1 + \frac{Z_1}{Z_M}\right)^2 - \left(1 - \frac{Z_1}{Z_M}\right)\right\} \\
&\times\exp\left(-2\sqrt{k_0^2\varepsilon_M(\omega) - k_\parallel^2}d_M\right)\frac{Z_M - Z_3}{Z_M + Z_3} \\
&+\left\{\exp\left(2\sqrt{k_0^2\varepsilon_2 - k_\parallel^2}d_I\right)\left(1 + \frac{Z_1}{Z_M}\right)^2 - \left(1 - \frac{Z_1}{Z_M}\right)^2\right\} \\
&\times\exp\left(2\sqrt{k_0^2\varepsilon_M(\omega) - k_\parallel^2}d_M\right)\frac{Z_1 + Z_M}{Z_1 - Z_M} = 0 \qquad (3.16)
\end{aligned}$$

where $k_\parallel$ is a wavenumber component parallel to the flat interfaces and Z_α ($\alpha = 1, 2, 3$, and M) is defined by

$$Z_\alpha = \frac{\sqrt{k_0^2\varepsilon_\alpha - k_\parallel^2}}{\varepsilon_\alpha}. \qquad (3.17)$$

For simplicity, the notation of k_0 is used, defined by $k_0 = \omega/c_0$. Note that the symbol Z_α has physical meaning such as

$$Z_\alpha = k_0 \times \text{impedance}.$$

In Eq. 3.16, the Z'_αs always appear as the ratios, which mean that the common factor k_0 cancels out in each Z_α term and that the Z'_αs work as impedance.[b]

Equation 3.16 provides the relation of $k_\parallel$ and ω. Practically, it is feasible to find numerical solutions $k_\parallel$ for a given ω because the value of $\varepsilon_M(\omega)$ is determined before finding $k_\parallel$. By iterating a loop for ω, we can obtain the $(k_\parallel, \omega)$ set to satisfy Eq. 3.16. The set provides dispersion curves of localized modes in the IMIMI structure of flat interfaces. Indeed, Eq. 3.16 was employed to evaluate the lowest mode in a concrete IMIMI structure in which semi-infinite I layers were air and SiO_2 (Iwanaga, 2011). Furthermore, a MIM structure free-standing in air was similarly analyzed to elucidate the underlying physics in the fishnet MM (Iwanaga, 2012b).

We note that the waveguide mode is not simply attributable to SPP(s) at the M–I interface(s). Indeed, the gradient $\partial\omega/(\partial k_\parallel)$ is smaller than that of the SPP Fig. 3.36c, implying the retardation in the waveguide of thickness d_I. Thus, the localized modes in the IMIMI structure, which are solutions of Eq. 3.16 and mainly distribute at the I layer of finite thickness, should be called plasmonic waveguide modes rather than gap plasmons because gap plasmon is an ambiguous term including local plasmons at metallic narrow gaps (Figs. 3.10 and 3.11).

3.2 Nanoparticle (NP) Plasmonic Resonators

In this section, metallic nanospheres, nanorods, and other nanoelements are described. The shapes are mathematically simply connected. Figure 3.5 displays transmission electron microscope (TEM) images of typical NP plasmonic resonators of spheres, pentagons, and triangles; the metallic nanostructures induce the blue-to-red colors in the dark-field images (Mock et al., 2002).

[b]Impedance is usually designated as Z. See Section 1.2.1 for the general definition.

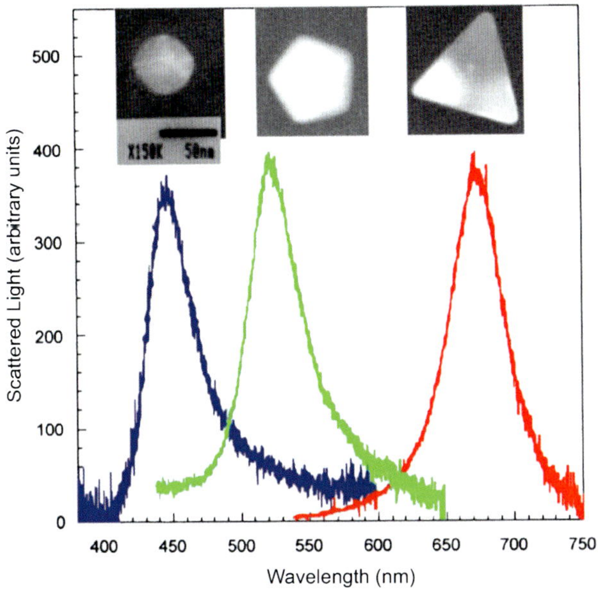

Figure 3.5 Metallic nanostructures and the dark-field scattering spectra. The spheres, pentagons, and triangles correspond to dark-field colors of blue, green, and red, respectively. The inset shows TEM images of the sphere, pentagon, and triangle from the left to right, respectively. Scale bar (black) in the TEM image indicates 50 nm. Reprinted with permission from Mock et al. (2002). Copyright 2002, AIP Publishing LLC.

Metallic NPs were used for coloration since long time ago in stained glass and wine bottles. The physical nature was first accounted for at the beginning of the 20th century (Maxwell-Garnett, 1904, 1906). When we consider a situation that metallic NPs are dispersed in a matrix of permittivity ε_{mat} at a low concentration of the filling ratio f, the effective permittivity ε_{eff} of the dispersed matrix is evaluated, using Lorentz model, such that

$$\frac{\varepsilon_{\text{eff}} - 1}{\varepsilon_{\text{eff}} + 2} = f\frac{\varepsilon_M - \varepsilon_{\text{mat}}}{\varepsilon_M + 2\varepsilon_{\text{mat}}} \tag{3.18}$$

where ε_{mat} and ε_M are permittivity of the matrix and the metal, respectively. By solving Eq. 3.18 for ε_{eff}, we obtain a relation for ε_{eff} such that

$$\varepsilon_{\text{eff}} = \varepsilon_{\text{mat}}\frac{\varepsilon_M(1 + 2f) + 2\varepsilon_{\text{mat}}(1 - f)}{\varepsilon_{\text{mat}}(2 + f) + \varepsilon_M(1 - f)}. \tag{3.19}$$

It is easily verified that $\varepsilon_{\text{eff}} \neq \varepsilon_{\text{mat}}$ and $\varepsilon_{\text{eff}} \neq \varepsilon_M$. Thus, Eq. 3.19 describes why the matrix of dispersed metallic NPs looks different in color from the nominally pure matrix and from the bulk metal. The Maxwell-Garnet model is often used as a handy model to evaluate permittivity of dispersed matrices.

Metallic NPs have been addressed in many books (for example, Kilmov, 2014), so we here do not go into the detailed results by theory and experiment. Instead, we outline Mie theory and briefly mention the main results.

3.2.1 *Mie Resonances in Metallic Spheres*

A rigorous theoretical result was reported by Mie in 1908 (Born and Wolk, 1999).[c] Originally, ideal conducting spheres were considered and nanospheres were not primary objects. Still, the theory is purely mathematical and applicable to the nanospheres of present interest.

Let us here describe the scattering configuration and theoretical implications for plasmonic resonators of metallic nanospheres. Figure 3.6 shows scattering configuration by a single metallic sphere in homogeneous medium (domain I). The sphere is assigned to domain II and the center is set to the coordinate origin. Incident wave with x polarization is set to travel along the z axis. The configuration is shown in the xz-section view. From the structure to analyze, it is suitable to formulate the electromagnetic (EM) wave scattering in the polar coordinate (r, θ, ϕ).

One of Maxwell equation in the Cartesian coordinate (Eq. 1.8) is written in the time-independent form, assuming that the temporal dependence is $\exp(-i\omega t)$:

$$\nabla \times \mathbf{H}(\mathbf{r}) = -i\omega\varepsilon_0\varepsilon(\omega)\mathbf{E}(\mathbf{r}) \tag{3.20}$$

where $\mathbf{r} = (x, y, z)$. Here let us set $k_1 = i\omega\varepsilon_0\varepsilon(\omega)$ and move to the spherical polar coordinate (r, θ, ϕ) defined in a standard manner such that[d]

---

[c]Consult the Section 14.5 for the detailed derivation.

[d]We here set $r = |\mathbf{r}|$.

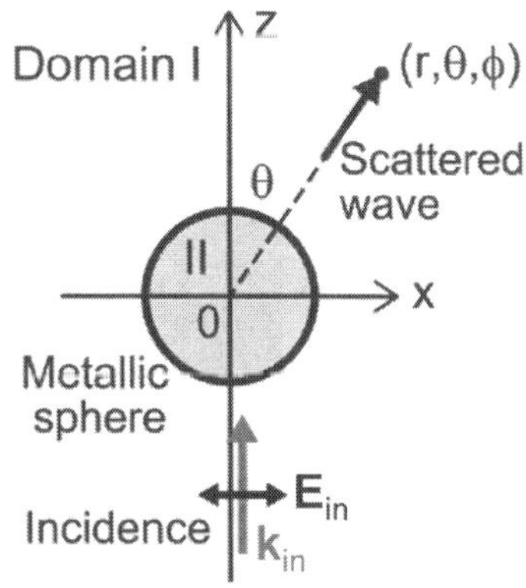

Figure 3.6 Schematic of scattering configuration by a metallic sphere (domain II) in homogeneous medium (domain I). The center of the sphere is set at the coordinate origin. The *xz*-section view is shown. Incidence of *x* polarization travels along the *z* axis.

$$x = r \sin\theta \cos\phi, \tag{3.21}$$

$$y = r \sin\theta \sin\phi, \tag{3.22}$$

$$z = r \cos\theta. \tag{3.23}$$

Equation 3.20 is expressed, using the polar coordinate, such that

$$-k_1 E_r = \frac{1}{r^2 \sin\theta} \left\{ \frac{\partial (r H_\phi \sin\theta)}{\partial \theta} - \frac{\partial (r H_\theta)}{\partial \phi} \right\}, \tag{3.24}$$

$$-k_1 E_\theta = \frac{1}{r \sin\theta} \left\{ \frac{\partial H_r}{\partial \phi} - \frac{\partial (r H_\phi \sin\theta)}{\partial r} \right\}, \tag{3.25}$$

$$-k_1 E_\phi = \frac{1}{r} \left\{ \frac{\partial (r H_\theta)}{\partial r} - \frac{\partial H_r}{\partial \theta} \right\}. \tag{3.26}$$

Another Maxwell equation Eq. 1.7 is expressed, similarly to Eq. 3.20, such as

$$\nabla \times \mathbf{E}(\mathbf{r}) = k_2 \mathbf{H}(\mathbf{r}), \tag{3.27}$$

using $k_2 = i\omega\mu_0$. Note that relative permeability μ is here set to $\mu = 1$ because non-magnetic metals are now under consideration mainly at optical frequencies. Equation 3.27 in the polar-coordinate representation is obtained by replacing $(-k_1, E, H)$ in Eqs. 3.24–3.26 with (k_2, H, E).

The solution of the scattering problem is written as superposition of two linearly polarized fields (${}^e\mathbf{E}$, ${}^e\mathbf{H}$) and (${}^m\mathbf{E}$, ${}^m\mathbf{H}$) such that

$$ {}^eE_r = E_r,\ {}^eH_r = 0, \tag{3.28} $$

$$ {}^mE_r = 0,\ {}^mH_r = H_r. \tag{3.29} $$

The suffices e and m denote electric and magnetic components along the radius, respectively. To resolve the set of Eqs. 3.24–3.26 and the replacing $[(-k_1, E, H) \rightarrow (k_2, H, E)]$ equations, substantial transformations using the Lagendre functions and so on are necessary, which can be verified in many other textbooks (for example, Born and Wolk, 1999).

We here look at representative results from the Mie theory. Figure 3.7 shows schematic illustrations of electric lines of force at a moment. The lines are projected onto the xz plane for clarity. The first to fourth resonances are shown in Fig. 3.7a–d, respectively. The nth resonance has $2n$ poles; half of them are positively (+) charged and the rest are negatively (−) charged. The poles oscillate from + to − and vice versa.

As an observed quantity, extinction cross sections of metallic spheres have been frequently reported. From the Mie theory, the

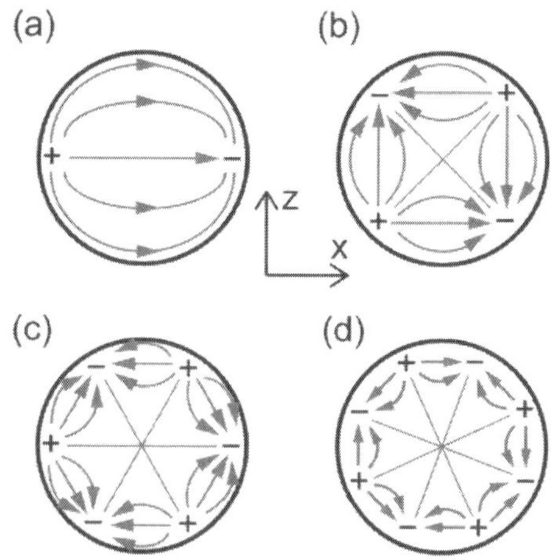

Figure 3.7 Schematics of the first to fourth resonant electric lines of force (arrows) at a moment in a metallic sphere. (a)–(d) correspond to the first to fourth resonances, respectively. They are projected onto the xz plane.

extinction cross section σ_{extinc} of the metallic sphere is given by

$$\sigma_{\text{extinc}} = \frac{\lambda^{(I)2}}{2\pi}\text{Re}\left\{\sum_{l=1}^{\infty}(-i)^{l+1}l(l+1)({}^{e}B_l + {}^{m}B_l)\right\} \quad (3.30)$$

where $\lambda^{(I)}$ is wavelength in the domain I, and ${}^{e}B_l$ and ${}^{m}B_l$ are coefficients characterizing the scattered wave at large distances from the metallic sphere.[e] The cross section has been experimentally confirmed, making the Mie theory to be an established theory for metallic spheres. Also, the Mie theory is now considered to be a basis to understand optical properties of metallic NPs.

3.2.2 *Modified Mie Resonances*

Metallic NPs were handled not only in size but also in shapes; an photoinduced experiment demonstrated that Au nanospheres were transformed into nanoprisms in solution (Jin et al., 2001). According to the increase in the variety in shapes of metallic NPs, full color scattering spectra in dark-field configurations were reported (Mock et al., 2002; Murray and Barnes, 2007). Thus, there was a requirement to extend the original Mie theory. To date, an empirical analysis was reported to use an analytic equation to describe experimental dark-field scattering spectra; the equation reproduced the scattering spectra of various NP shapes in the region beyond Rayleigh approximation (Kuwata et al., 2003).

3.2.3 *Planar Array of Metallic Disks*

Figure 3.8a shows a schematic of planar array of Ag nanodisks on quartz substrate. In Fig. 3.8b, a typical reflectance (R), transmittance (T), and absorbance (A) spectra are shown with red dotted, blue dashed, and black solid lines, respectively. The linear optical spectra were numerically calculated by the RCWA+S-matrix method (see Section 1.5.1) The Ag disks were set to be isolatedly arrayed with periodicity of 200 nm, diameter (d) of 50 nm, and height (h) of

[e] The concrete expressions of ${}^{e}B_l$ and ${}^{m}B_l$ are rational equations including the Hankel functions.

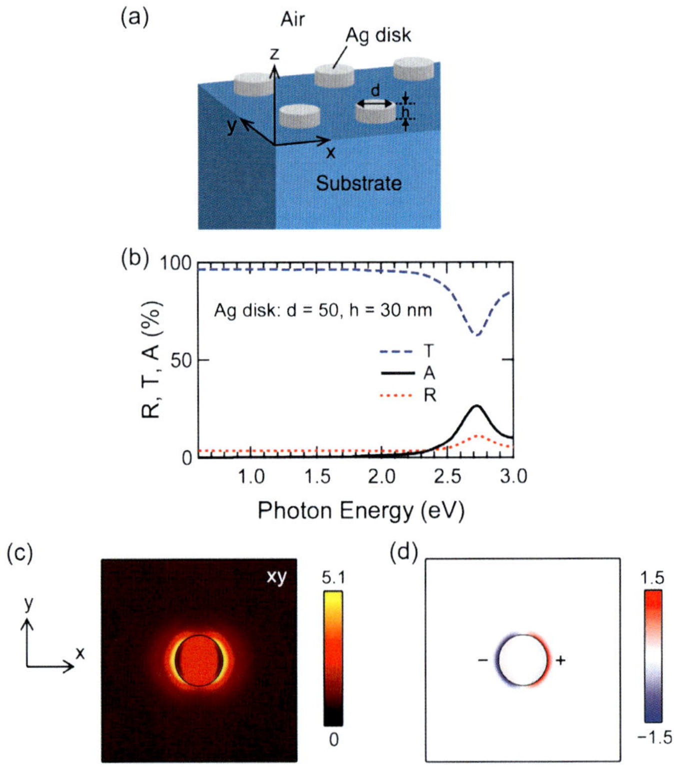

Figure 3.8 (a) A schematic of planar array of metallic nanodisks. The periodicity was set to 200 nm. (b) R, T, and A spectra of array of Ag disks (d = 50 and h = 30 nm), at the normal incidence from the air side, which are shown with red dotted, blue dashed, and black solid lines, respectively. (c) |**E**| distribution at the absorption peak of 2.7 eV in (b). (d) A snapshot of E_z component, corresponding to (c). The *xy* section was taken at the top of the Ag disk. The incident |**E**| was set to 1.0.

30 nm. The *x*-polarized incidence was set to travel from the air side. A prominent absorption peak appears at 2.7 eV.

To visualize the resonant mode at 2.7 eV, the corresponding |**E**| distribution was also calculated by the RCWA+S-matrix method. Figure 3.8c shows the |**E**| distribution on a *xy* section located at the top of the Ag disk. The E-field magnitude is strongly localized at the edge of the Ag disk, suggesting that the resonance originates from a Mie-type local plasmon in each Ag disk. The x-polarized incidence was set to be normal to the xy plane and $|\mathbf{E}| = 1.0$.

Figure 3.8d shows a snapshot of E_z component, corresponding to Fig. 3.8c and being calculated similarly. Note that $\mathrm{Re}(E_z)$ is plotted as the E_z component. Obviously, the E_z-field orientations are opposite at the edges of the Ag disk. Note that the direction of the incident polarization was set to x polarization. Still, the E_z component appears at the side wall of the Ag disk, which means that longitudinal component is substantially induced around the metallic nanostructure.

Recently, the electron energy-loss spectrum of truncated Ag nanospheres on a thin Si_3N_4 substrate were examined (Li et al., 2015). It was observed that the Ag spheres of 50 nm diameter exhibit the lowest peak at 2.777 eV. The experimental configuration is quite similar to that in Fig. 3.8a. The result is in good agreement with the numerical result that shows a peak at 2.7 eV in Fig. 3.8b. The slight shift probably comes from detailed difference in the structure including the configuration and partly from the classical limit assuming the local response of the materials (see Section 4.2).

3.3 NP-Assembled Plasmonic Resonators

Metallic nanoelements described in the previous Section 3.2 were often produced by chemical synthesis. Furthermore, the self-assembled in chemical synthesis significantly stimulated the studies based on the metallic nanoelements (Prodan et al., 2003; Fan et al., 2010). Thus, many chemists have contributed to the research on assembled metallic NPs.

On the other hand, nanoantennas such as disk-pair dimers, bowtie antenna, nanorod, and rod-pair antenna have been some of the most preferred plasmonic resonators so far (Novotny and Hecht, 2012; Giannini et al., 2011). This is probably because the resonant property is feasible to understand. In the structures, the hot spots appear at a gap between the metallic nanostructures.

Figure 3.9 illustrates NP-assembled structures, which can be fabricated on substrates by the top-down nanofabrication methods (see Section 1.6); (a) dimer, (b) tetramer (or quadrumer), and

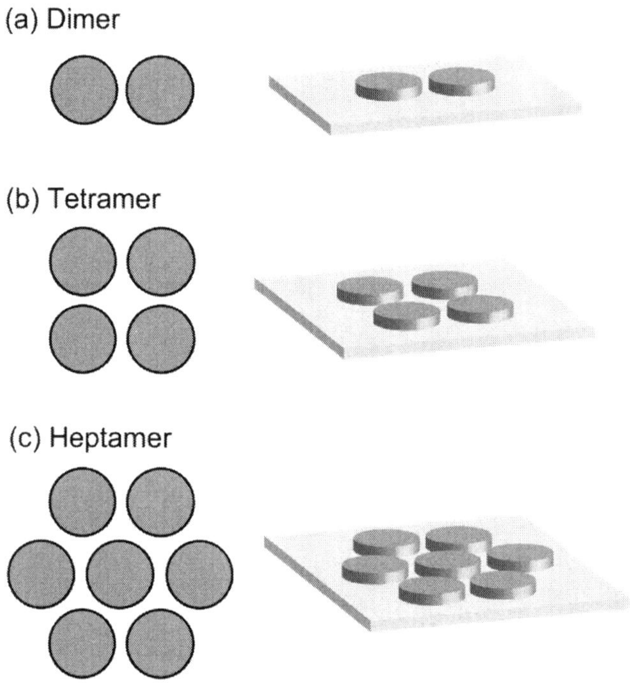

Figure 3.9 NP-assembled metallic nanostructures, which can be fabricated by top-down nanofabrication methods. (a) Dimer. (b) Tetramer. (c) Heptamer. Top-view structures are illustrated at the left-hand side and the 3D views are shown at the right.

(c) heptamer. The top views are illustrated at the left-hand side in Fig. 3.9 and the 3D views are shown at the right-hand side. The assembled NPs were so often Au in the experiment so far (Hentschel et al., 2010). Each structure was fabricated on a transparent substrate such as silica. In particular, large E-field intensity enhancement at the narrow gaps was expected in the experiment such as plasmon–exciton strong coupling and plasmon-enhanced Raman scattering; see Sections 5.5 and 5.6 for the details, respectively.

In this section, we survey typical properties of NP-assembled plasmonic resonators. Gap plasmons resident in the plasmonic resonators are mainly addressed.

3.3.1 *Gap Plasmons and Nanoantenna*

Gap plasmons appear at nanometer space in plasmonic resonators of assembled metallic NPs. Here let us describe representative features induced in the assembled systems.

Dimers of gap 1 nm Figure 3.10 shows gap plasmon resident in 1 nm metallic gap. The unit structure comprises two circular Ag disks and is periodically arrayed as illustrated in Fig. 3.10a. The diameter of the Ag disk was set to 50 nm and the periodicity of the array was 200 nm. The height of the disk along the *z* axis was set to 30 nm. Each Ag-disk pair was set to be distant enough to respond independently to illuminating EM waves. Substrate was set to be quartz of refractive index of 1.46.

Figure 3.10b shows numerically calculated R, T, and A spectra under normal incidence of *x* polarization from the air side; accordingly, the A is written as A_x. The first resonance appears at 2.09 eV and the second resonance at 2.69 eV. In the diffraction-free range, the absorbance A was defined by

$$\mathrm{A} = 100 - \mathrm{R} - \mathrm{T}. \tag{3.31}$$

Figures 3.10c and 3.10d present the $|\mathbf{E}|$ distributions on *xy* sections at the first (2.09 eV) and second (2.69 eV) resonances, respectively. The *xy* sections were taken at the top and bottom of the Ag-disk pairs in Figs. 3.10c and 3.10d, respectively. The central areas are enlarged below, as indicated by gray arrows. For the incident E-field magnitude of $|\mathbf{E}| = 1.0$, the first resonance has intense E-field magnitude at the 1 nm gap, whereas the second has four highly enhanced $|\mathbf{E}|$ distributions distant from the 1 nm gap. Such enhanced $|\mathbf{E}|$ distributions are often called *hot spots*.

Figure 3.10e shows *xy*-section $|\mathbf{E}|$ distribution of the first resonance at the bottom of the Ag-disk pair, as indicated in Fig. 3.10f by a gray dotted arrow. The E-field magnitude at the bottom is distinct from that at the top Fig. 3.10c and the most enhanced $|\mathbf{E}|$ appears in a highly localized manner at the Ag-disk edges opposite to the 1 nm gap. The $|\mathbf{E}|$ distributions on the first resonance at 2.09 eV at are quite different at the top Fig. 3.10c and bottom Fig. 3.10e of the Ag-disk.

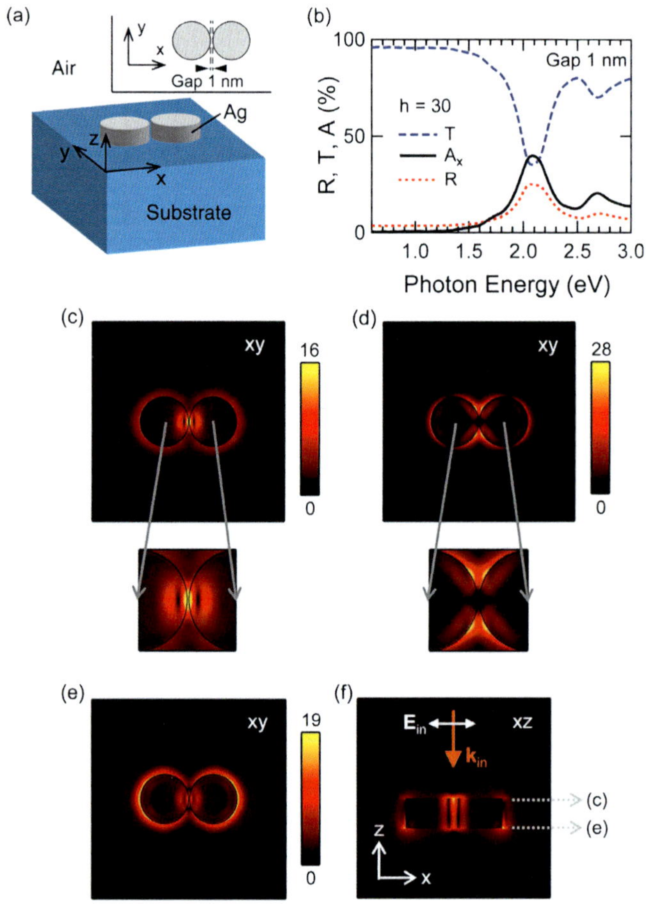

Figure 3.10 (a) Schematic of Ag-disk dimer with gap 1 nm. (b) R, T, and A spectra of periodic array of the Ag-disk dimer under the x-polarized normal incidence. (c) and (d) $|\mathbf{E}|$ distributions at the first (2.09 eV) and second (2.69 eV) resonances, respectively. (e) The $|\mathbf{E}|$ distribution at the bottom on the first resonance. (f) The $|\mathbf{E}|$ distribution at the first resonance on the *xz* section through the center of the Ag-disk pair. The positions of the *xy* sections in (c) and (e) are indicated by dotted arrows. $|\mathbf{E}_{\text{in}}| = 1.0$.

Figure 3.10f displays an *xz* section of the $|\mathbf{E}|$ distribution at the first resonance; the section was taken at the center of the Ag-disk pair. It was found that the hot spots appear in the 1 nm gap at the top of Ag-disk pairs [position (c)] and at the edges opposite to the gap at the bottom [position (e)]. Thus, even in the thin dimer, the hot spots distribute in the different way at the top and bottom, suggesting that the air and substrate sides should be carefully examined in actual experiment using the hot spots. Also, the |**E**| distributions on the first and second resonances at the bottom are qualitatively distinct from each other.

Dimers of gap 0 nm Figure 3.11a is a schematic illustration of array of two-Ag-disk contacting structure. The gap between the disk was set to 0 nm in the structure of Fig. 3.10. The term, gap plasmon, sounds like that nm gap is necessary. However, gap plasmons similar to those in Fig. 3.10 were found even in the 0 nm gap structure.

Figure 3.11b shows numerically calculated R, T, and A_x spectra at the normal incidence with x polarization. Two resonances appear; the first resonance is located at 1.30 eV and the second at 2.42 eV.

For the incident y polarization perpendicular to the disk-to-disk direction, the A_y spectrum shown with gray solid line has a peak at 2.75 eV, which is close to the peak of isolated Ag disk in Fig. 3.8; this result strongly suggests that the resonance is independent of the gap and dominantly comes from the Ag disks themselves.

Figure 3.11c shows xy-section |**E**| distribution of the first resonance; the part is enlarged as indicated by gray arrows. Hot spots emerge around the 0 nm gap in a separate way and are highly enhanced such that max(|**E**|) = 26 where the incident |**E**| was set to 1.0. Figure 3.11d shows that the second resonance has four hot spots, which appears in a qualitatively similar manner to the second resonance in Fig. 3.10d. The xy sections of the |**E**| distributions in Figs. 3.11c and 3.11d were taken at the bottom of the Ag-disk pairs.

In the cases of the gap 0 and 1 nm in Figs. 3.10 and 3.11, the first resonance is directly affected whether or not the gap exists, whereas the second resonance show similar |**E**| distributions to each other.

Related to the gap of the Ag dimers, Fig. 3.12 shows change of absorbance spectra under x polarization, depending on the gap distances from 0 to 10 nm. The cases of 0 and 1 nm were shown again (Figs. 3.10 and 3.11). Obviously, as the gap distances becomes larger, the lowest absorption peaks moves to higher energies. In the large gap limit, the Ag-disk configuration becomes equivalent to the array of isolated Ag disk in Fig. 3.8 where the peak is located at

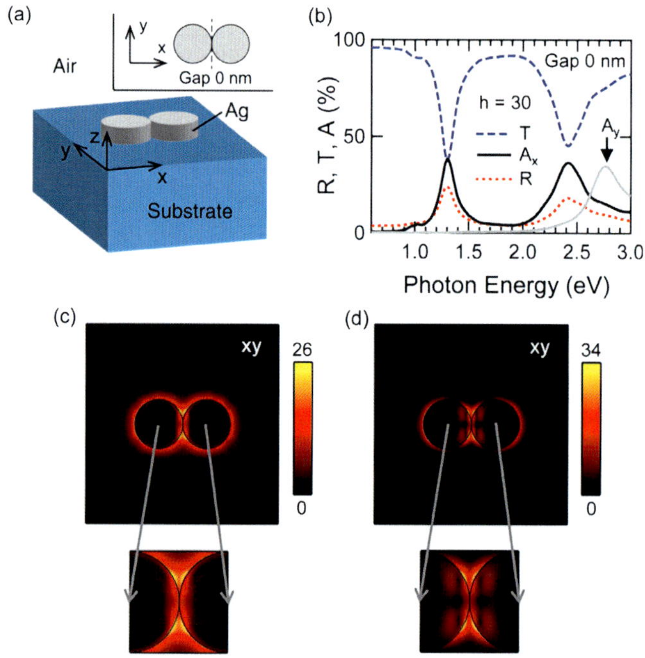

Figure 3.11 (a) Schematic of a gapless 50 nm-diameter Ag-disk dimer of 30 nm thickness. The periodicity of the dimer array was set to 200 nm. (b) Optical spectra of periodic array of Ag-disk pair with gap 0 nm. R, T, and A spectra at the normal incidence are shown with red, blue and gray lines, respectively. (c) and (d) $|\mathbf{E}|$ distributions at the first (1.30 eV) and second (2.42 eV) resonances, respectively. The *xy* sections were taken at the bottom of the contacting dimer. The central areas are enlarged below, as indicated by gray arrows. The incidence satisfied $|\mathbf{E}_{\text{in}}| = 1.0$.

2.7 eV. Thus, remarkable effect of the dimerization is observed in the cases that the gap distances are 0 and 1 nm.

Anisotropic light absorption of the dimers Figure 3.13 describes light absorption dependent on incident angles. Although the angle-dependent absorption has not been usually paid attention, a prominent property appears as follows. Figure 3.13a is a schematic of optical configuration. The Ag-disk pairs of 1 nm gap have the A_x peak at 2.09 eV (Fig. 3.12) and are excited by the incidence of $\mathbf{k}^{(xz)}$ (red arrow) with p polarization (i.e., $\mathbf{E}_{\text{in}} \| xz$ plane) and of $\mathbf{k}^{(yz)}$

of 500×500 nm^2 dimension. The position of Ag nanorod is indicated by the rectangular shape. The E_x, E_z, and H_y components are the snapshots and their local polarity is indicated by $\pm$. We particularly note the following points.

(i) As is clearly seen, the EM field distributions of the Ag nanorod are not affected by the nearest neighbor nanorods; therefore, the EM fields primarily represent the plasmonic resonance resident in the single Ag-nanorod antenna.
(ii) Although the incidence polarization is x-polarized, the E_z component is the major E-field component and the set of (E_z, H_y) characterizes the resonant EM field. Such E_z distributions were experimentally measured by scanning near-field optical microscope (for example, Fig. 3.20).
(iii) The magnitude of E field, $|\mathbf{E}|$, shows that it is qualitatively in agreement with the E_z distribution and moreover is largely enhanced, taking the maximum of 18 for the incident $|\mathbf{E}_{\text{in}}| = 1.0$. The resonant $|\mathbf{E}|$ in the nanoantenna is quite larger than that in Ag disk (Fig. 3.8c).
(iv) From the comparison of $|\mathbf{E}|$ with E_x and E_z, it turns out that the imaginary part of the E field takes large values.
(v) As for the magnetic field, the H_y component shows that the distribution corresponds to the E_z (not E_x), forming the lowest-order standing-wave distribution, and is also found to be substantially enhanced. The incidence was set to satisfy $|\mathbf{H}| = 1.0$.

The extinction spectra and EM field distributions were calculated by the RCWA+S-matrix method.

3.3.2 *Hybridized Plasmons*

The notion of plasmon hybridization was first introduced for assembled metallic core-shell structures (Prodan et al., 2003). The scheme for chemical bonding of molecules was imported to plasmonics and worked well to describe energies of the assembled plasmonic resonators.

Figure 3.16 schematically illustrates energy diagram for hybridized plasmons in a Ag-sphere dimer. The Ag spheres (gray) are

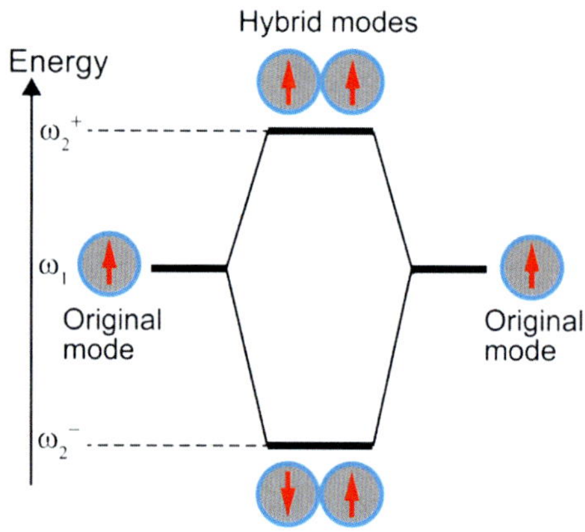

Figure 3.16 Schematic energy diagram for plasmon hybridization. Ag spheres (gray) with thin insulator shell (pale blue) are taken as example. Resonant oscillations at a moment are designated with arrows.

assumed to be coated by thin insulator shell (pale blue) to prevent direct contact of the spheres. Original mode in a single Ag sphere at energy $\hbar\omega_1$ is here assumed to be the first mode in Fig. 3.7a. By coupling the two original modes, the eigenenergy splits into two levels of $\hbar\omega_2^-$ and $\hbar\omega_2^+$, and two hybrid modes are formed. From the analogy of two-level coupling, it has been confirmed that anti-parallel and parallel oscillations correspond to the lower and higher energy levels, respectively.

This hybridization model is true for metallic nanorod coupling systems; indeed, Fig. 2.7a is consistent with the present model. Assembled metallic nanodisks such as oligomers (Hentschel et al., 2010) and nanosphere clusters (Fan et al., 2010) were also studied along the scheme above. The hybridization model is moreover applied later to a more complex system of heteroplasmon coupling (Fig. 3.46).

3.4 Single-Layer Lattices

Thanks to the development of nanofabrication technology, periodic structures of plasmonic resonators are available in daily research activities. The periodic structures are often referred to as plasmonic crystals (PlCs) though the periodic structures are mostly restricted to 2D or 1D structures with finite depth.

In this section, we address single-layer lattices comprised more complicated plasmonic resonators than the NPs in Sections 3.2 and 3.3. Even when we restrict plasmonic resonators to the single-layer ones, there is already a large number of reports on them. Readers can find the examples in a recent review addressing the single-layer (or on-top) plasmonic resonators (Carminati et al., 2015).

3.4.1 *Periodically Perforated Metallic Films*

Simple hole array As described in Chapter 1, air-hole array in metallic thick film attracted great interest as extraordinary transmission (EOT) (Ebbesen et al., 1998) and stimulated many studies (Genet and Ebbesen, 2007; Garcia-Vidal et al., 2010). A typical T spectrum is shown in Fig. 1.15. The numerous studies confirmed that light can be modified in the spatial distributions and transmit even through tiny holes compared to the wavelength in vacuum. The modification is possible due to the resonances originated from SPPs. Nowadays, there is no surprise in the effect; from a historical viewpoint, we now understand that the effect was widely recognized by the observation of the *extraordinary* transmission.

Annular hole array Coaxial metallic nanostructures attracted interest in connection to EOT (Baida and VanLabeke, 2003; Fan et al., 2005; Poujet et al., 2007). Through the studies on the transmission origin, it became clear that the coaxial structures allow plasmonic waveguide modes inside the annular hole and the resonances were classified based on the waveguide modes (Baida et al., 2006; de Waele et al., 2009; Catrysse and Fan, 2009).

In the coaxial structures, there is a series of waveguide modes. As for TE[f] polarization that is excited by linear polarization in the section of coaxial structures, cut-off wavelength for the lower *m*th

[f]Transverse electric.

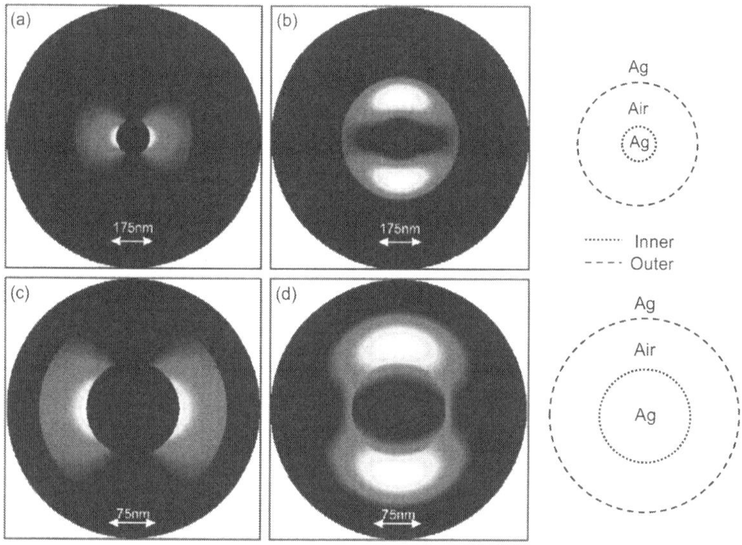

Figure 3.17 E-field intensity $|\mathbf{E}|^2$ in single coaxial plasmonic waveguides, made of Ag and air, at the cut-off wavelengths λ_{m1}^{c}. (a) and (b) The first and second modes for $R_{\text{out}} = 250$ nm, respectively. (c) and (d) The first and second modes for $R_{\text{out}} = 150$ nm, respectively. The inner radius R_{in} was fixed at 75 nm in (a)–(d). The inner (dotted line) and outer circles (dashed line) were shown at the right-hand side for guides to the eye. (a) to (d) were reprinted with permission from Baida et al. (2006). Copyright 2006 by the American Physical Society.

mode λ_{m1}^{c} is approximated by

$$\lambda_{m1}^{c} \approx \frac{\pi(R_{\text{out}} + R_{\text{in}})}{m} \tag{3.33}$$

where R_{out} and R_{in} denote the outer and inner radius, respectively (Baida et al., 2006). The cut-off wavelengths are defined at the crossing points of the dispersion of waveguides modes and the light cone in air. Note that the TE modes are generally indexed such as TE_{mn}, the lowest mode is TE_{11}, and TE_{mn} ($n > 1$) modes appear at far shorter wavelengths, not contributing here.

Figure 3.17 shows numerically calculated $|\mathbf{E}|^2$ distributions of plasmonic waveguide modes in the coaxial waveguides, which were studied as isolated single Ag waveguides (not periodic array). Figure 3.17a and 3.17b show the first and second waveguide modes, respectively, for $R_{\text{out}} = 250$ and $R_{\text{in}} = 75$ nm; the wavelengths were set to the cut-off wavelengths λ^{c}_{m1} ($m = 1, 2$). The first mode is strongly localized around the inner Ag rod whereas the second mode mainly distributes in the air domain between the inner and outer Ag walls.

Figures 3.17c and 3.17d are similar to Figs. 3.17a and 3.17b, except for $R_{\text{out}} = 150$ nm; note that the scale is changed. As for the appearance, note that the inner radius is enlarged as compared to Figs. 3.17a and 3.17b. The $|\mathbf{E}|^2$ distributions in the coaxial structure of $R_{\text{out}} = 150$ nm are also similar to those in Figs. 3.17a and 3.17b. Thus, it is naturally inferred that the mth plasmonic waveguide modes have qualitatively similar EM-field distributions. The numerical calculations in Fig. 3.17 were implemented using FDTD method and used Drude-type permittivity for Ag (Baida et al., 2006).

From the analysis for single coaxial waveguides, high T is expected employing the waveguide modes in the periodic array of coaxial waveguides. Indeed, several papers reported measured efficient T (Fan et al., 2005; Poujet et al., 2007; de Waele et al., 2009; Iwanaga et al., 2012). Let us next look into a typical experimental result on periodic array of metallic coaxial structure.

Figure 3.18a shows a top-view SEM image (upper panel) of an annular PlC, which was fabricated in Al film on SiO_2 substrate. The section at the white line in the SEM image is illustrated at lower panel. Figure 3.18b shows optical configuration for T measurement. Figure 3.18c displays measured T spectra at incident angle θ from $\theta = 0^\circ$ to 40° under p polarization. Note that, although the annular aperture in the unit cell is 2.8%, T at 1200 nm and $\theta = 0^\circ$ (arrow) exceeds 20%, exhibiting EOT. Figure 3.18d shows numerically calculated T spectra, which correspond to the measured T spectra in Fig. 3.18c and present good agreement. Both T spectra are shown with offset for clarity. Plasmonic resonances in the annular PlC were clarified into localized modes in the annular holes and diffraction-associated modes that sensitively depend on incident angles, based on the EM field distributions and the dispersion in $(k_{\parallel}, \omega)$ plane where $k_{\parallel}$ is wavenumber projected onto the xy plane in Fig. 3.18b (Iwanaga et al., 2012).

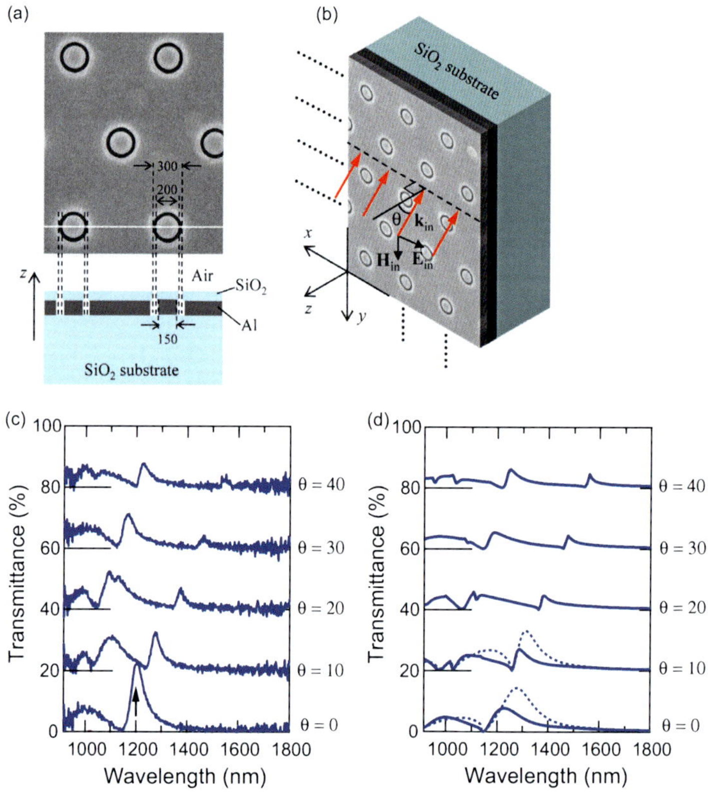

Figure 3.18 Coaxial Al PlC on SiO_2 substrate. (a) Top-view SEM image (upper) and the dimension of in-depth profile (lower). The periodicity was 900 nm. (b) 3D-view schematic. (c) Measured T spectra at p polarization. (d) Calculated T spectra by the RCWA+S-Matrix method. Adapted with permission from Iwanaga et al. (2012). Copyrighted by the American Physical Society.

Related to the annular-hole array structures, we note that off-center coaxial structures have larger enhanced E-field intensity at the narrower gaps (Hao et al., 2008) and that negative refractive index was numerically evaluated by filling the annular structure with semiconductor (Burgos et al., 2010), in which the subwavelength array of the annular structures was regarded as a MM.

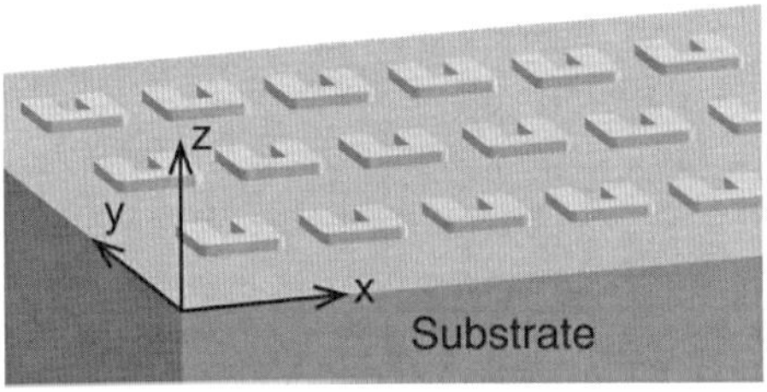

Figure 3.19 Schematic illustration of an array of U-shaped SRRs for optical frequencies.

3.4.2 *Array of Plasmonic Resonators*

U-shaped SRR array Figure 3.19 shows a schematic of U-shaped SRR array on a substrate. The structure was conceived by simplifying the double SRR at microwave (Fig. 1.17). Although the double SRRs were made at THz (Yen et al., 2004), the structure is demanding at optical frequencies because of the structural complexity, so that the simplified SRRs were introduced for optical frequencies (Linden et al., 2004; Enkrich et al., 2005). Analysis using effective permittivity and permeability resulted in magnetic responses at the SRRs in Fig. 3.19 and stimulated MMs research at optical frequencies. If one assumes that the U-shaped SRR is a single coil, z-component of magnetic field is expected to induce current in the coil. Note that the magnetic field is not static but oscillates at optical frequencies. The magnetic field of nonzero z component is applied to the SRR array under s-polarized oblique incidence; for example, such a configuration is realized using y-polarized incidence that is oblique in the xz plane. Indeed, the magnetic response was reported at the s polarization.

Figure 3.20 shows a set of results on near-field measurement of U-shaped SRR array (Zentgraf et al., 2008). Figures 3.20a and 3.20b display atomic force microscope (AFM) images of U-shaped SRR arrays, which have scale bars of 640 nm in both panels; in Fig. 3.20a, the SRRs were arrayed in a dense and closely packing way. Figure 3.20c shows simulated R spectra of the two SRR arrays: solid red line denotes the R spectrum of the array of small SRRs, including the first and third resonances shown with symbols $1^{(S)}$ and $3^{(S)}$,

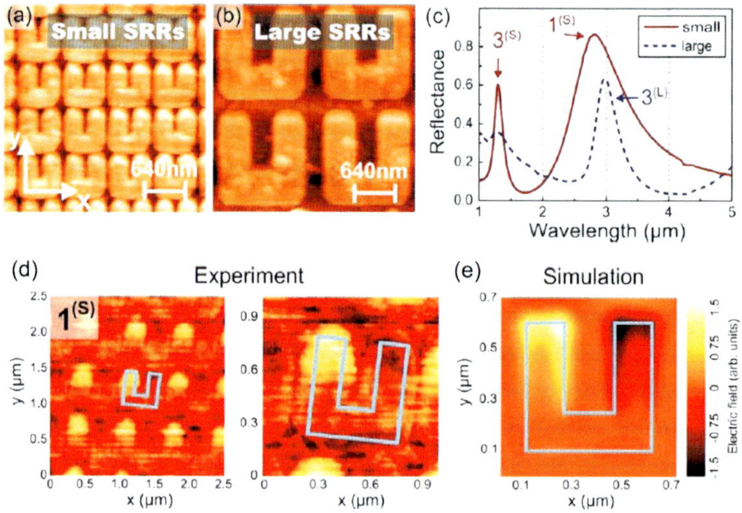

Figure 3.20 (a) and (b) AFM images of array of U-shaped small and large SRRs, respectively. (c) Far-field R spectra. (d) E-field images measured by an apertureless SNOM. (e) Numerically calculated E-field, corresponding to (d). Adapted with permission from Zentgraf et al. (2008).

respectively; dashed blue line represents the R spectrum of the array of large SRRs, including the third resonance shown with symbol $3^{(L)}$.

Figure 3.20d shows scanning near-field optical microscope (SNOM) images of the small SRR array on the first resonance. The SNOM used apertureless probe tip; the typical setup is similar to Fig. 5.20. The apertureless SNOM images show the *z* component of E-fields perpendicular to the U-shaped SRR at a fixed phase of the E field, where the *z* axis is defined to be perpendicular to the *xy* plane in Fig. 3.20a. The positions of a small SRR are indicated by pale blue lines.

Figure 3.20e shows a simulated E_z image corresponding to Fig. 3.20d. A reasonable agreement of the measured data with the simulation was obtained, suggesting that, in each SRR, the lowest electric-dipole oscillation takes place as for the E_z component indicating $\lambda/2$ distribution of polarity $+-$, just like Fig. 3.15d. The lowest oscillation was also confirmed in an unfolded metallic rod (i.e., a straight metallic rod) using the SNOM technique. Based on the result on the SRR in Fig. 3.20d, it is naturally inferred that the mode named $3^{(L)}$ exhibits the third-order multipole oscillation on the E_z component indicating $3\lambda/2$ distribution of polarity $+-+-$, which was confirmed similarly to the mode $1^{(S)}$.

3.5 Collective Oscillation Associated with Longitudinal Component in Plasmonic Resonators

Plasmons themselves are collective excitation of conductive electrons in metals. However, if we regard plasmon as an elementary excitation, the collective states are possible to realize. In this section, we survey collective plasmon states (Iwanaga, 2012a), which have not been widely recognized. We also discuss the relation of the collective states with the longitudinal component of plasmonic resonances, described in Chapter 2.

Figure 3.21 shows schematics of metallic nanorod arrays: (a) free-standing Ag-rod array on SiO_2 substrate and (b) Ag-rod array embedded in Si layer on SiO_2 substrate. The rectangular Ag rods are arrayed along the $(-1, 1)$ direction in the *xy* plane and are set to contact at the edges. Let us next examine the resonant modes.

Figure 3.21b shows a 3D-view schematic of array of rectangular Ag nanorods. Figure 3.22a shows the unit cell in the *xy* plane. The size of the unit cell is 250×250 nm^2 and the unit cell consists of Ag and Si. Each nanorod has 50×50 nm^2 in the *xy* section and is set to contact at the edge with each other. Figure 3.22b presents

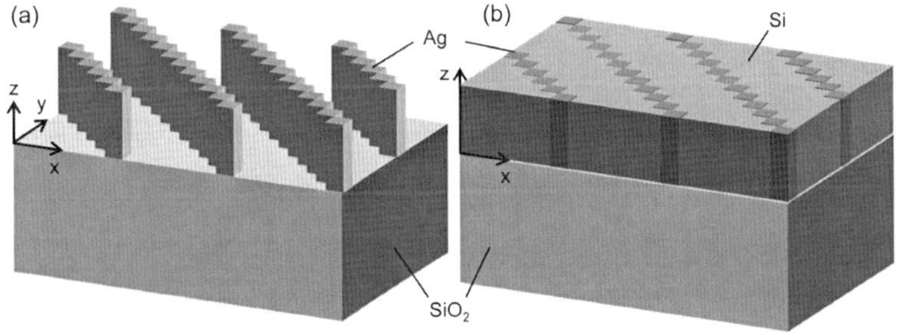

Figure 3.21 Schematic of rectangular nanorod array. (a) Nanorod array standing in air on SiO_2 substrate. (b) Nanorod array embedding in Si on SiO_2 substrate. Adapted from Iwanaga (2012a).

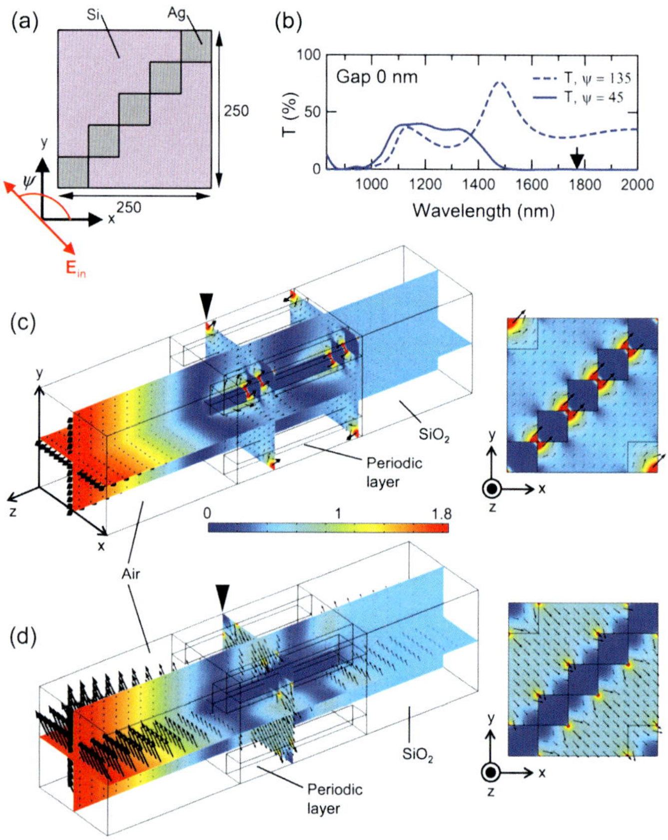

Figure 3.22 Collective plasmons in rectangular Ag nanorod array. (a) *xy* section of the nanorod array in periodic layer. (b) T spectra at two perpendicular polarizations of $\psi = 45^\circ$ and 135°. (c) Collective plasmonic resonance with polarization parallel to the nanorod array ($\psi = 45^\circ$). Left: 3D view; right: *xy* section view. The section is indicated by black cone in the left. (d) Non-collective plasmons with polarization perpendicular to the array ($\psi = 135^\circ$), represented similarly to (c). Adapted from Iwanaga (2012a).

numerically calculated polarized T spectra at $\psi = 45^\circ$ and 135°; the ψ is the angle of the x axis and the incident polarization $\mathbf{E}_{\text{in}}$ in Fig. 3.22a. The T spectra exhibits strong dependence on the polarization at the wavelength range longer than 1500 nm; the T at $\psi = 135^\circ$ keeps several tens % while the T at $\psi =$

45° takes almost 0%. To clarify the phenomena underlying the strong polarization dependence, let us next examine the E-field distributions at 1780 nm, indicated by arrow.

Figures 3.22c and 3.22d show snapshots of E-field at 1780 nm for $\psi = 45^\circ$ and 135°, respectively, which were calculated by a FEM-based software (see Section 1.5.3). At the left-hand side, the 3D-view E-field distributions are displayed. The domain was taken to contain the unit cell. At the right-hand side, the E-field distributions of a *xy*-section view are shown; the section positions are indicated by cones at the 3D views. The absolute value of the E field is shown with color and the E-field vectors are with arrows.

At $\psi = 45^\circ$, the E field is strongly localized around the Ag-nanorods contacting edge of Ag nanorods, which is shown with red. This is the origin of the 0% T in Fig. 3.22b. The E-field localized state is a light-trapping state.

On the other hand, at $\psi = 135^\circ$, there is not trapped states; instead, hot spots are seen at the corner of the Ag nanorods. The hot spots are insensitive to incident wavelength and are induced at a continuum. This kind of hot spots are less recognized in general but contributes as optical loss in wide wavelength range. Actually, the T at $\psi = 135^\circ$ is tens % and associated with tens % optical loss. In plasmonic structures, it is necessary to keep in mind that this kind of optical loss (or light absorption) is inevitable.

Role of longitudinal component Phenomenological observations are noted just above. Let us discuss the results from the transverse and longitudinal components (see Sections 2.2 and 2.4). The collective mode appeared at $\lambda > 1500$ nm and $\psi = 45^\circ$ and the hot spots at $\psi = 135^\circ$ are excited in continuous manners. There was not any prominent T peak for the continuous resonant EM-field distributions; therefore, the resonant behaviors do not appear as discrete dispersion diagram. These properties are features of the longitudinal component, which are dispersionless and continuous.

As noted in Section 2.2, the longitudinal component is enhanced associated with the resonant transverse components. In this sense, the longitudinal and transverse components are not independent to each other; rather, they are simultaneously excited in plasmonic resonators. Therefore, the hot spots are seen in an enhanced way;

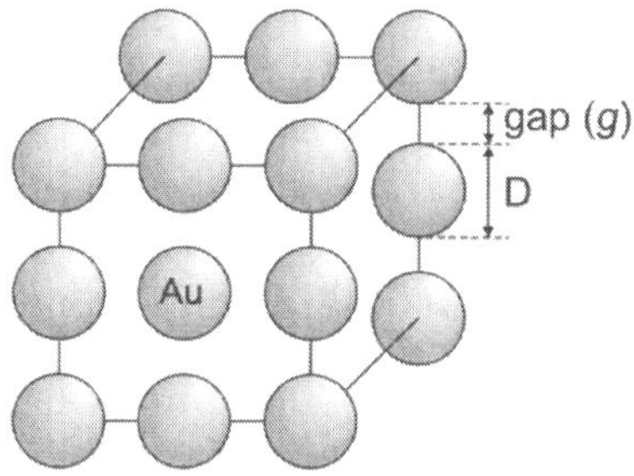

Figure 3.23 3D array of Au spheres with gap distance (g). Each Au sphere has the diameter (D) and are arrayed in a cubic manner.

however, they are continuously (or almost everywhere) induced at a wide wavelength range. The longitudinal components are observed as non-zero backgrounds in the absorption spectra.

3.6 Plasmonic Resonators of Simply Stacked Structures

3.6.1 *Three-Dimensional (3D) Stacking Array of Metallic Spheres*

Even contemporary nanofabrication techniques do not easily reach 3D structures with nm precision. Still, we can examine them in numerical ways. Let us first consider 3D array of Au nanospheres as an example of simply stacking structures.

Figure 3.23 shows a schematic illustration of 3D array of Au spheres in a cubic manner. The Au spheres are assumed to have the diameter D and arrayed with the nearest-neighbor gap (g). The array resembles so-called colloidal crystals in the appearance; the colloidal crystals are usually made of polystyrene, opal, and semiconductor spheres and produced in self-assembled ways (Born, 2013). To date, although there seems no report on the fabrication on Au-sphere array, the production would be possible in principle.

Figure 3.24a shows single-layer Au-nanosphere array set in a quartz cell filled with water. The periodic array was set to spread in the xy plane. The diameter D of Au spheres was set to 100 nm. Figure 3.24b shows extinction spectra for several g from $g = 4$ to 100 nm. The extinction was defined by Eq. 3.32 for the T represented at the

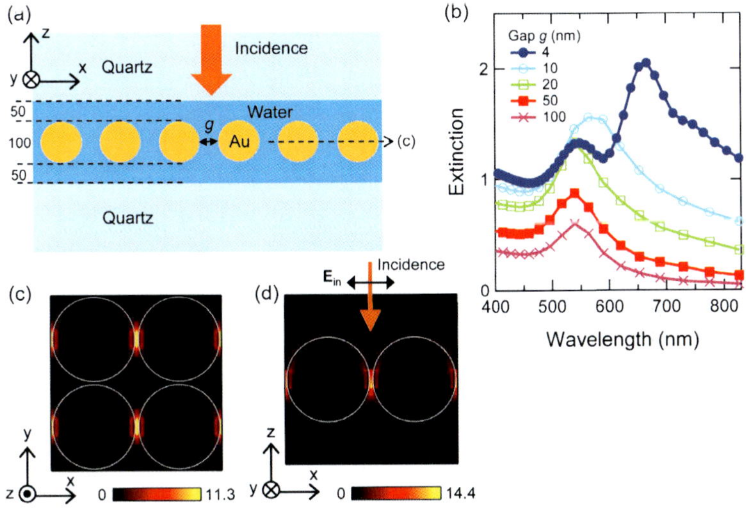

Figure 3.24 (a) Schematic of periodic array of Au spheres in water, drawn in the units of nm. (b) Extinction spectra of periodic array of metallic spheres, Gap distances were varied from 4 to 100 nm. (c) *xy*-section-view $|\mathbf{E}|$ distribution of the Au-sphere array of 4 nm gap. The z position is indicated by a dashed-line arrow in (a). (d) *xz*-section-view $|\mathbf{E}|$ distribution, corresponding to (c). The $|\mathbf{E}|$ distributions in (c) and (d) were induced at 666 nm and $|\mathbf{E}_{\text{in}}| = 1.0$.

0–1 range (not %). Incident polarization was set to be x polarization. When each Au sphere is isolated such as $g = 100$ nm (purple line with crosses), the extinction peak energy is due to the bulk material property (i.e., permittivity) of Au. The extinction spectra are similar to those for $g = 50$ (red line with closed squares) and 20 nm (green line with open squares); the increase in the extinction comes from the increase in Au density in the unit cell. As the g decreases below 10 nm (light blue with open circles), the extinction peaks exhibit red shift. When the g is 4 nm (blue line with closed circles), the most prominent peak appears at 666 nm and are expected to come from a new plasmonic resonance at the narrow gaps between the Au spheres; besides, there appears a small peak at 550 nm that has the same origin to the peaks in cases of $g \geq 20$ nm.

Figure 3.24c displays $|\mathbf{E}|$ distribution in the single-layer Au-sphere array of gap $g = 4$ nm, representing the resonance at the extinction peak of 666 nm. The *xy* section was taken at the centers of Au spheres, as indicated in Fig. 3.24a. Figure 3.24d shows $|\mathbf{E}|$ distribution in the *xz* section through the centers of Au spheres. Because the incident polarization was *x*-polarized, the intense $|\mathbf{E}|$ distributions were induced at the gaps across the *x* axis, being more than 10-fold enhanced for the incidence of $|\mathbf{E}| = 1.0$. Thus, the extinction peak at 666 nm in the case of $g = 4$ nm is ascribed to gap plasmon. The extinction spectra and E-field magnitude were calculated by the RCWA+S-matrix method. The refractive index of water was set to a representative value of 1.33.

Figure 3.25a shows a schematic of two-layer Au-sphere array with periodicity $100 + g$ nm. The spheres are assumed to be periodically and infinitely arrayed in the *xy* plane. The diameter of

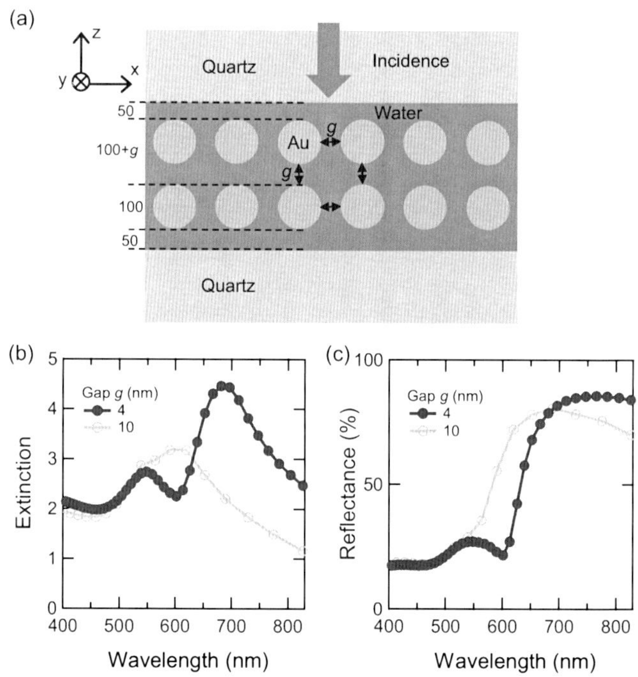

Figure 3.25 (a) Schematic of array of Au spheres that is cubic in the *xy* plane and stacked along the *z* axis. (b) and (c) Calculated extinction and R spectra of periodic array of metallic spheres, respectively. Gap distances were set at 4 (line with closed circles) and 10 nm (line with open circles).

each Au sphere was set to 100 nm. The array, which is a stacking structure of the single-layer structure in Fig. 3.24, is also assumed to be in water between the quartz walls.

Figure 3.25b shows extinction spectra for $g = 4$ (line with closed circles) and 10 nm (line with open circles). The extinction peaks appear at similar wavelengths to those of single-layer array, and the values of extinction are approximately double. The results mean that the number of gaps between the Au spheres increases twice and the excited gap plasmons also become twice. Thus, the stacking of Au-sphere array is useful to assemble the gap plasmon resonators.

Figure 3.25c shows R spectra corresponding to the extinction spectra in Fig. 3.25b. At longer wavelengths than 700 nm, the R qualitatively becomes large and close to the R of bulk Au. The extinction peaks appear at the wavelengths where the R begins decreasing with moving shorter wavelengths.

3.6.2 *Stratified Metal–Insulator Metamaterials (SMIMs)*

Brief overview SMIMs comprise one of the simplest unit cell, that is, metal–insulator multilayers of thin films, as illustrated in Fig. 3.26, which shows six-layer unit cell as an example. Of course, the simplest

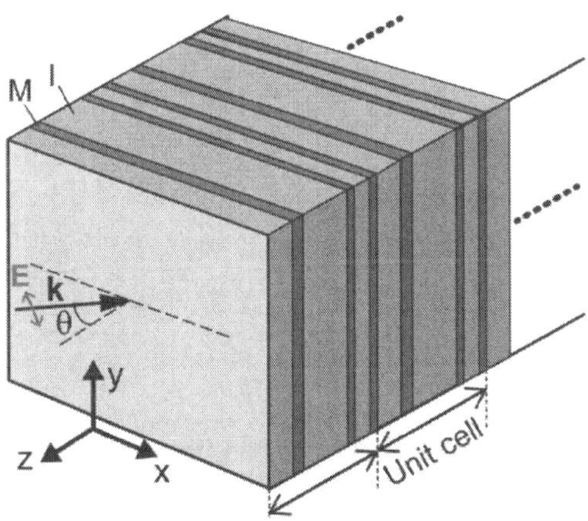

Figure 3.26 A schematic illustration of a SMIM. As an example, a SMIM of six-layer unit cell is drawn. Adapted from Iwanaga (2015).

unit cells comprise of two layers, that is, one metal and one insulator layers.

In structural viewpoint, SMIMs include MIM structure addressed in Section 3.1 and therefore are considered to inherit some of the properties. Resonances originating from the waveguide modes are expected and indeed appear later (Section 3.6.5).

SMIMs attracted interest in 1990s from the viewpoint of the transmission window in spite of including an opaque component (i.e., metal), and then they were studied in the context of photonic crystals (Scalora et al., 1998). Currently, the unintuitive transmission window is qualitatively understood as a simple optical property in the low-loss range appearing above the effective plasma frequency; this feature is described later (Fig. 3.33b).

At the transmission window, nonlinear responses also attracted interest. Indeed, highly enhanced third-order nonlinearity was theoretically predicted at 650 nm for multilayers of unit (Cu 16 nm/SiO_2 100 nm) (Bennink et al., 1999), and second and third harmonic generations were studied experimentally (O'Donnell and Torre, 2005) and theoretically (Scalora et al., 2010).

Since the theory on super-resolution (SR) imaging using a single Ag layer (Pendry, 2000), much attention has been paid to another aspect of stacked metal–insulator multilayers as media realizing SR imaging. Indeed, SMIMs exhibit prominent features such as diffraction-suppressing transmission mode (Ramakrishna and Pendry, 2003; Salandrino and Engheta, 2006; Jacob et al., 2006) and negative refraction (Verhagen et al., 2010; Xu et al., 2013). The diffraction-suppression mode was employed to realize hyperlens as described later (see Section 3.6.3).

In most studies, the SMIMs were analyzed assuming effective media model, in which the dispersion equation is expressed as

$$\omega^2 = \frac{k_\perp^2}{\varepsilon_\parallel} + \frac{{k_\parallel}^2}{\varepsilon_\perp} \tag{3.34}$$

where $\varepsilon_\parallel$ and $\varepsilon_\perp$ are components in effective permittivity tensor for uniaxial material, respectively, and the symbols $\parallel$ and $\perp$ denote parallel and perpendicular to the principal axis, respectively. When a relation such as $\varepsilon_\perp < 0 < \varepsilon_\parallel$ holds, Eq. 3.34 means hyperbolic dispersion in the $(k_\parallel, k_\perp)$ plane; therefore, SMIMs are often called

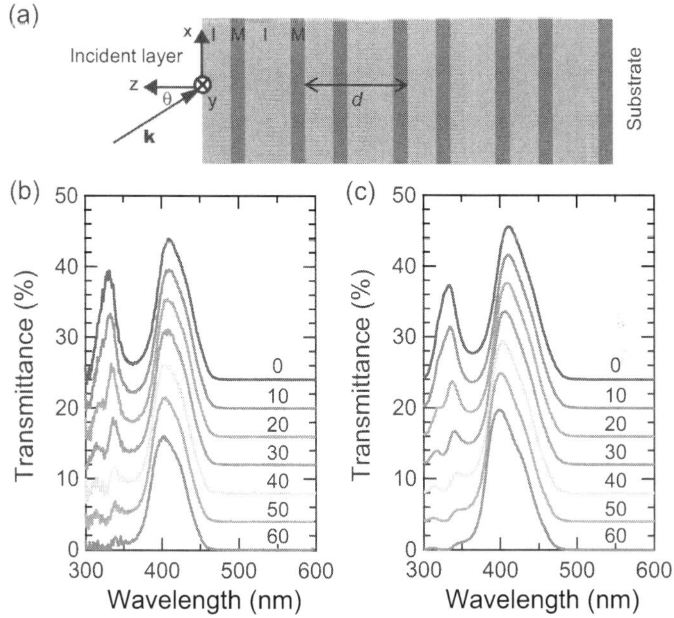

Figure 3.27 Optical properties of SMIMs. (a) Schematic of a SMIM of four-layer unit cell, which contains four unit cells (i.e.,16 layers) in total. (b) Measured T spectra under p polarization and incident angles from 0° to 60°. The spectra are shown with offset. (c) Numerically calculated T spectra, corresponding to (b). Adapted from Iwanaga (2015).

hyperbolic MMs (as reviews, Poddubny et al., 2013; Drachev et al., 2013). The model makes the understanding of SMIMs easy, whereas there is a limitation for the model (Iwanaga, 2007) and the model is always needed to be justified.

Basic properties of SMIMs The most obvious feature is the SMIMs have efficient transmission window despite they include an opaque element of metal. Figure 3.27 shows the illustrative example (Iwanaga, 2015).

Figure 3.27a illustrates a SMIM in a *xz*-section view where the principal axis was set to be *z* axis, and shows that the SMIM has the unit cell of four layers. The thickness of IMIM layers in the unit cell was set to 50, 25, 80, and 25 nm from left to right, respectively; then, the periodicity d was 180 nm. The M was chosen to be Ag and the I to be SiO_2.

Figure 3.27b shows measured T spectra of the SMIM of the four-layer unit cell. The incidence was p-polarized (i.e., $\mathbf{E}_{\text{in}} \| xz$ plane in Fig. 3.27a), and the incident angles were varied from 0° to 60°. The T spectra are displayed with offset for clarity. The SMIM has transmission windows at 330 and 410 nm. The window at 330 nm stems from the feature of Ag itself; due to the small imaginary part of the permittivity, Ag thin films generally exhibits T peak at 330 nm. In contrast, the transmission window at 410 nm comes from the stacked structure in the SMIM. Although the total metallic thickness in the SMIM is $200 (= 25 \times 8)$ nm, the T reaches approximately 20%. In addition, note that the window hardly changes the wavelength range under p polarization, suggesting that the resonance responsible for the T window is qualitatively distinct from SPP-origin resonances that are sensitive to incident angles. The unit cell was particularly designed to realize the transmission window at 400–420 nm, which was employed in SR imaging as shown later (Fig. 3.31).

The specimen measured for Fig. 3.27b was made with ion-beam sputtering that enable to precisely control the rate of deposition of Ag and SiO_2 such as 0.31 ± 0.03 and 0.17 ± 0.03 nm/s, respectively. At present, other film-making methods are unlikely to produce the multilayers of Ag and SiO_2 in the highly precise manner.

Figure 3.27c shows numerically calculated T spectra, corresponding to the measurement above. Evidently, the calculated T spectra well reproduce the measured ones. Thus we can state that good-quality SMIMs are in good agreement with the numerical implications.

Photonic Bloch states in SMIMs Multilayer structures without any optical loss are characterized using the photonic Bloch states (Yeh et al., 1977), which are eigenmodes in the periodic structures; the photonic Bloch states are purely Hermite state, similar to the Bloch states of periodic systems in quantum mechanics. However, the SMIMs are multilayer structures with optical loss originating from the opaque metal of finite imaginary part of permittivity. Therefore, in order to apply the photonic-Bloch-state method to the SMIMs, it is inevitable to introduce some extended treatment. Note that, unless any modification is introduced, the results from the Bloch-state analysis are, in principle, hard to be justified. Here we briefly describe an extended Bloch-state (or eigenmode) analysis (Iwanaga, 2012c).

The photonic Bloch states are derived from 2×2 transfer matrix $T_{\rm mat}$ for a unit cell of SMIMs. Generally, the $T_{\rm mat}$ describes boost across the unit cell such that

$$T_{\rm mat} \begin{pmatrix} F^{(+)}(z) \\ F^{(-)}(z) \end{pmatrix} = \begin{pmatrix} F^{(+)}(z+d) \\ F^{(-)}(z+d) \end{pmatrix} \tag{3.35}$$

where $F^{(\pm)}$ are coordinate components of EM wave propagating for $\pm z$ directions, respectively. A typical configuration is depicted in Fig. 3.27a. Some technical details are elaborated as follows.

(i) We can generally assume that the EM wave propagates in the xz plane, that is, the wavevector $\mathbf{k} \| xz$ plane; then, the field components $F^{(\pm)}$ are the y components.
(ii) There are two possible polarizations, that is, TE and TM polarizations. For TE polarization, $F^{(\pm)} = E_y^{(\pm)}$, and, for TM polarization, $F^{(\pm)} = H_y^{(\pm)}$.
(iii) The Bloch states are independent of the choice of z. However, the position z has to be explicitly determined which layer include the z. Here we set z to be $z - 0$, which means that the z is in an I layer in Fig. 3.27a.

The concept of Bloch states is that periodic systems are characterized using phase factors associated with the periodicity. Accordingly, the boost of EM-field vector at the right-hand side of Eq. 3.35 is replaced with a phase factor $\exp(iK_z d)$; the K_z is an index characterizing the Bloch states. We thus reach an eigenvalue equation such that

$$T_{\rm mat} \begin{pmatrix} F^{(+)}(z) \\ F^{(-)}(z) \end{pmatrix} = \exp(iK_z d) \begin{pmatrix} F^{(+)}(z) \\ F^{(-)}(z) \end{pmatrix}. \tag{3.36}$$

The 2×2 $T_{\rm mat}$ has two eigenvalues; one of them represents an EM wave propagating for $+z$ direction and the other for $-z$ direction. The two K_z is uniquely determined to exclude exponentially growing solution.

For the case in accordance with the both-ends arrow in Fig. 3.27a, the T_{mat} is explicitly expressed as

$$T_{\text{mat}} = \mathcal{B}_{\text{I}}(d_4)\mathcal{I}_{\text{IM}}\mathcal{B}_{\text{M}}(d_3)\mathcal{I}_{\text{MI}}\mathcal{B}_{\text{I}}(d_2)\mathcal{I}_{\text{IM}}\mathcal{B}_{\text{M}}(d_1)\mathcal{I}_{\text{MI}} \tag{3.37}$$

where $\mathcal{B}$ and $\mathcal{I}$ denote boost and interface matrices, respectively, the subscripts M and I indicate M and I layers, respectively, and d_i ($i = 1$ to 4) stands for thickness of each layer.[g] As we mentioned above [the remark (iii)], the z is set at $z - 0$ in an I layer.

Recalling the matrix-form boundary condition (Eq. 1.19) and the modification (Eq. 1.20), the interface matrix $\mathcal{I}_{\text{MI}}$ for TE polarization has concrete form such that

$$\mathcal{I}_{\text{MI}} = \begin{pmatrix} 1 & 1 \\ k_z^{(\text{M})}/\varepsilon_{\text{M}} & -k_z^{(\text{M})}/\varepsilon_{\text{M}} \end{pmatrix}^{-1} \begin{pmatrix} 1 & 1 \\ k_z^{(\text{I})}/\varepsilon_{\text{I}} & -k_z^{(\text{I})}/\varepsilon_{\text{I}} \end{pmatrix}. \tag{3.38}$$

Another interface matrix $\mathcal{I}_{\text{IM}}$ is readily obtained by exchanging M and I in Eq. 3.38. Note that, for TM polarization, the relative permittivity ε is replaced with relative permeability μ.

The z component of wavevector is related to the x component such that

$$k_z^{(\alpha)} = \sqrt{\varepsilon_\alpha \mu_\alpha \frac{\omega^2}{c_0^2} - \left\{k_x^{(\alpha)}\right\}^2} \tag{3.39}$$

where α represents M or I. Equation 3.39 is equivalent to dispersion equation in the material α.

Boost matrix $\mathcal{B}$ is written as

$$\mathcal{B}_\alpha(d_j) = \begin{pmatrix} \exp\left(ik_z^{(\alpha)}d_j\right) & 0 \\ 0 & \exp\left(-ik_z^{(\alpha)}d_j\right) \end{pmatrix}. \tag{3.40}$$

Using the explicit expression of the boost and interface matrices, the determinant of T_{mat} is exactly equal to unity. It follows from

$$\det\left\{\mathcal{B}_\alpha(d_j)\right\} = 1 \tag{3.41}$$

$$\det(AB) = \det(A)\det(B) \tag{3.42}$$

$$\det\left(A^{-1}\right) = \det(A)^{-1} \tag{3.43}$$

for arbitrary matrices A and B where $\det(A) \neq 0$. Note that the relation

$$\det(T_{\text{mat}}) = 1 \tag{3.44}$$

[g]In more general cases, T_{mat} for the N-layer unit cell is obtained by repeatedly multiplying the set of $\mathcal{B}_{\text{I}}\mathcal{I}_{\text{IM}}\mathcal{B}_{\text{M}}\mathcal{I}_{\text{MI}}$.

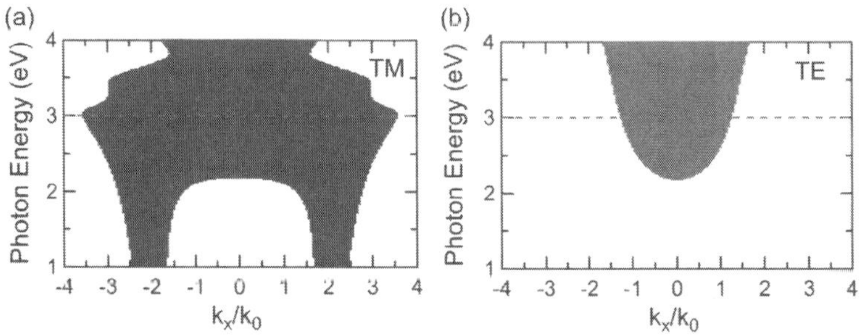

Figure 3.28 Photonic Bloch bands of the SMIM employed in the SR imaging. The unit cell of the SMIM comprises four layers and is same to the unit cell in Fig. 3.27a. (a) TM polarization and (b) TE polarization. The quasi-transmission condition was set to Eq. 3.48.

holds irrespective of the constituent materials M and I. Equation 3.44 is rewritten as

$$\exp\left\{iK_z^{(1)}d\right\}\exp\left\{iK_z^{(2)}d\right\} = 1, \tag{3.45}$$

using the two eigenvalues in Eq. 3.36, i.e., $K_z^{(j)}$ $(j = 1, 2)$. Equation 3.45 means that one of the two eigenvalues always satisfies

$$\left|\exp\left\{iK_z^{(j)}d\right\}\right| \leq 1 \tag{3.46}$$

where $j = 1$ or 2. The Bloch state satisfying Eq. 3.46 is non-divergent solution for the $+z$-direction propagation. The other eigenvalue corresponds to the solution for $-z$-direction propagation. Note that, for purely loss-less cases, the eigenvalues are exactly equal to unity in Eq. 3.46 and represent Hermite states (Yeh et al., 1977).

Now we can visualize the Bloch states for SMIMs. The propagating states is, strictly, quasi-transmission states, which are quantitatively characterized. For example, a condition that

$$0 \leq \text{Im}(K_z) < 0.1 \tag{3.47}$$

determines relatively low-loss states, enabling quite long-range propagation (Iwanaga, 2015). This relatively low-loss states are consistent with the T peaks in Figs. 3.27b and 3.27c.

Figure 3.28 shows photonic Bloch bands of the SMIM defined in Fig. 3.27a, which has the unit cell of (I 50/M 25/I 80/M 25) in the units of nm, and represents a condition that

$$0 \leq \mathrm{Im}(K_z) < 1. \tag{3.48}$$

This condition on K_z allows relatively large loss for the quasi-transmission states. The materials I and M were set to be SiO_2 and Ag, respectively.

The polarizations in Fig. 3.28 were as follows: (a) TM polarization and (b) TE polarization. The horizontal axis is k_x component normalized by k_0 (k_0: wavenumber of light in vacuum) and the vertical axis is photon energy in eV. Because the T_{mat} includes the k_x component through Eq. 3.39, the eigenvalues K_z implicitly depends on the k_x. The plots in Fig. 3.28 were obtained by finding the K_z satisfying Eq. 3.48 for each k_x. The Bloch bands enable to visualize some features of the SMIM:

(i) The Bloch bands are essentially polarization-dependent. The TE band is similar to those of dielectric that is fairly transparent above 2.2 eV, whereas the TM band has waveguide-mode-origin band below 2.2 eV and k_x-wide band above 2.2 eV. Note that the light cone exists at approximately ± 1.5 in the plot of Fig. 3.28.

(ii) The k_x-wide band for TM polarization is useful to realize SR imaging employing SMIMs. Indeed, the state at 3.0 eV, indicated by dashed line in Fig. 3.28, is employed in an experimental SR imaging, as described later (Figs. 3.30–3.32).

3.6.3 *Application of SMIMs (1): Hyperlens*

Proof-of-principles experiment on hyperlens Figure 3.29a shows a schematic of experimental configuration for SR imaging through a cylindrical-shaped hyperlens (Liu et al., 2007). The role of the hyperlens, which comprises SMIM on cylindrical mold, is to transfer original image inside the cylindrical shape to the outer surface in a magnified way.

Figure 3.29b shows a SEM image (left) of the original shape that was prepared in a size under the diffraction limit and the corresponding optical image through the hyperlens (right); the two separated distance was about 180 nm in the original image and the working wavelength λ was 365 nm. Thus, the obtained optical

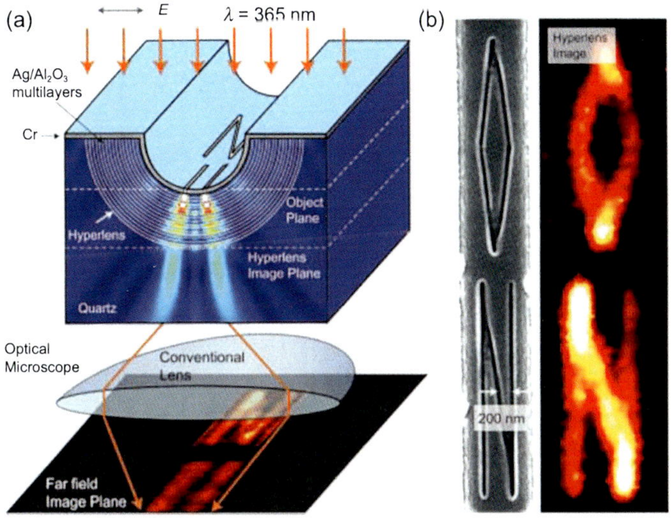

Figure 3.29 SR imaging by a cylindrical-shaped hyperlens. (a) Schematic of the structure and configuration. (b) SEM image of the original shape (left) and the optical image through the hyperlens. From Liu et al. (2007). Reprinted with permission from AAAS.

image was better than a typical diffraction limit of $\lambda/2$. On the outer surface, the optical image was magnified to the size over diffraction-limit and observed in an ordinary optical imaging technique. The SMIM consisted of Ag/Al_2O_3 multilayers and had 16 layers in total.

The experimental demonstrations in Fig. 3.29 and (Smolyaninov et al., 2007) were on the way indicated by theoretical suggestions (Salandrino and Engheta, 2006; Jacob et al., 2006). An extended experiment using semi-sphere hyperlens was also conducted to make images of 2D objects (Rho et al., 2010), in a similar way to Fig. 3.29. The experiments confirmed the principles of the image transfer by hyperlens in SR manners whereas the demonstrations were limited to the objects that were directly made on the hyperlens, so that it has not been clear whether it is possible to incorporate the hyperlens into actual optical microscopes. A recent progress in this point is described next.

Realistic simulation on SR imaging Figure 3.30 shows SR-imaging test using the SMIM designed for visible wavelength only using Ag

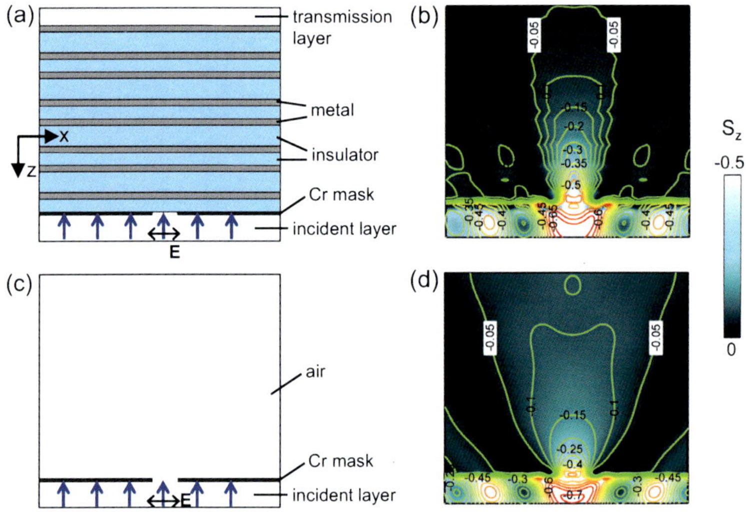

Figure 3.30 SR-imaging test with/without the SMIM. (a) SR test configuration, composed of a SMIM (M = Ag, I = SiO_2) and a deep subwavelength Cr slit. (b) S_z component, representing propagation through the Cr slit in (a). $|S_z|$ is plotted with pseudo-color and contours (green lines). (c) Reference configuration for (a). (d) S_z component in (c), plotted similarly to (b). Adapted with permission from Iwanaga (2014). Copyright 2014, AIP Publishing LLC.

and SiO_2, whose basic properties were described in Fig. 3.27. Figure 3.30a shows a configuration for the SR test. A Cr mask of 23 nm thickness was assumed to produce a subwavelength image; 100 nm width slit was located at the center in Fig. 3.30a and perpendicular to the *xz* plane. Incidence of 410 nm corresponding to the dashed line in Fig. 3.28a was set to be normal to the Cr mask and the polarization was *x*-polarized. The incident and transmitted layers were assumed to be air.

Figure 3.30b shows EM-field distribution in the configuration of Fig. 3.30a. In the subwavelength domain, EM wave propagation through the Cr slit is characterized by the *z* component of Poynting vector S_z, which is plotted in Fig. 3.30b. The EM wave that reaches the transmission layer is dominantly composed of far-field component and therefore proportional to the E-field intensity. The S_z contour of -0.05 means that the intensity is 5% for the incidence and forms a detectable image; the contour has the lateral width of 143 nm at the interface of the SMIM and the transmission layer, indicating that the diffraction (i.e., divergence of the image) is substantially suppressed.

Figure 3.30c depicts a reference configuration to Fig. 3.30a, removing the SMIM. Figure 3.30d shows the S_z component, plotted in a similar way to Fig. 3.30b. Without the SMIM, diffraction of the image takes place as it is expected. The lateral width of the image was approximately 800 nm, indicating that the subwavelength image is lost due to the diffraction limit.

Thus, Fig. 3.30 shows that the SMIM works as diffraction-suppressing medium and supports the validity of the Bloch state analysis in Fig. 3.28.

Implementation of HLA to optical microscope Figure 3.31a shows a schematic of hyperlens array (HLA) (Iwanaga, 2014). Objects are assumed to be located near the inner surface, and the transferred and magnified images are observed on the outer surface. In the dotted rectangular, the details of the structure are illustrated; gray denotes Ag and pale blue SiO_2.

Figure 3.31b presents a SEM image taken from the bottom side in Fig. 3.31a. The HLA comprises hexagonal array of hyperlens, whose periodicity was 6 μm. The diameter of each hyperlens was approximately 1.3 μm. The shape of hyperlens was close to semi-sphere. The inset shows a section-view SEM image of a hyperlens; the white scale bar indicates 1 μm.

Figure 3.31c illustrates HLA-implemented microscope. Basically, the HLA was introduced in a dark-field optical microscope. The picked up images by the objective lens go into a CMOS (complementary metal oxide semiconductor) camera with the maximum 100 frame/s. Incident light travels through a total reflection prism, and evanescent field induced by the total reflection is scattered by objects. Practically, the top surface of the prism induces weak scattering due to the slight surface roughness, emitting far-field illumination light in an incoherent way. The efficiency of the far-field scattering was empirically estimated to be 10^{-4} and was probably higher than the direct scattering of the evanescent field by objects. Therefore, the limit of spatial resolution, L, in the over-diffraction-

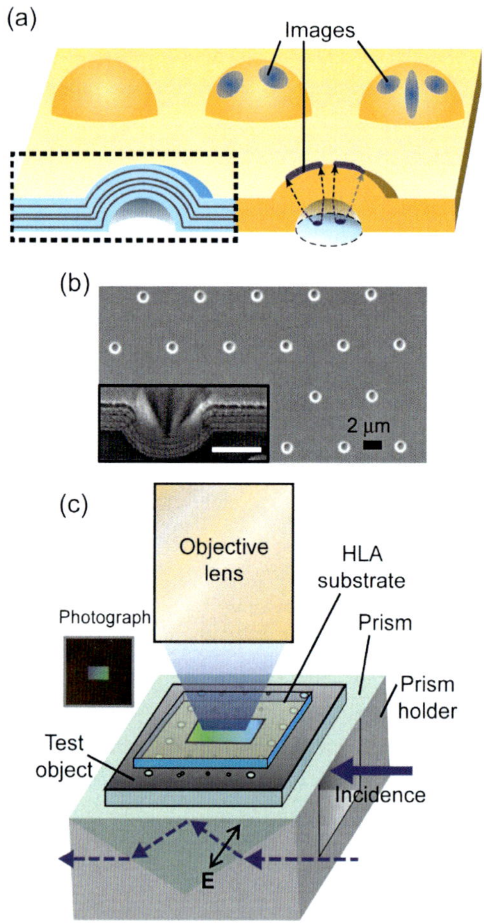

Figure 3.31 HLA-implemented microscope. (a) Schematic of HLA. (b) SEM image of HLA, taken from the bottom side in (a). The inset shows a section-view of a hyperlens. (c) Setup of HLA-implemented microscope, which is based on a dark-field microscope using a total reflection prism. The magnified images are observed from the top. Reprinted with permission from Iwanaga (2014).

limit regime is evaluated for the incoherent illumination by

$$L = \frac{0.61\lambda}{\text{NA}} \tag{3.49}$$

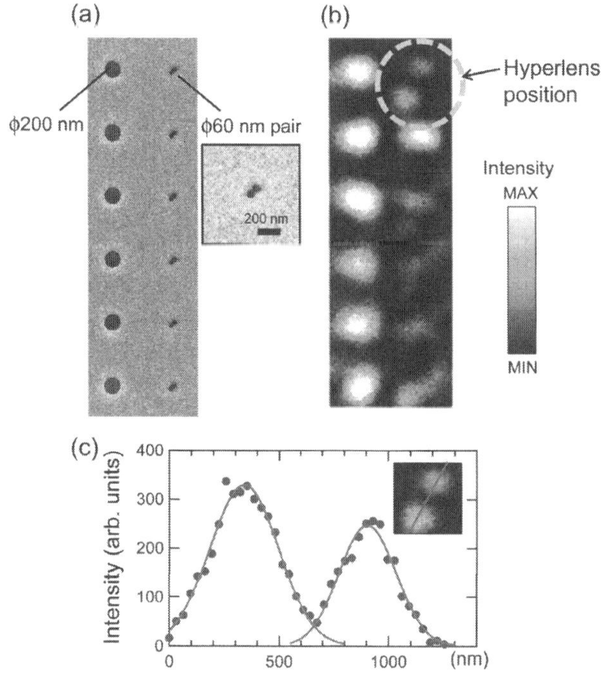

Figure 3.32 SR imaging by the HLA-implemented microscope. (a) SEM image of test objects. The inset magnified a pair of subwavelength air holes introduced in a Cr film. The center-to-center distance was 71 nm. (b) Optical image of the test object. The observed intensity is represented with pseudo-color. The dashed-line circle indicates position of a hyperlens. (c) A section profile of the test-object image. The data are shown with dots. The inset indicates the position of the section. Reprinted with permission from Iwanaga (2014).

where λ is wavelegth of illumination light and NA denotes numerical aperture of the objective lens (Born and Wolk, 1999).[h]

Figure 3.32 shows a set of results on SR imaging by the HLA-implemented microscope. Figure 3.32a shows a SEM image of test objects, which were set periodically at periodicity of 1 μm. The left-hand column was an array of air holes of diameter 200 nm introduced in a Cr film of 23 nm thickness deposited on a quartz substrate. The right-hand column was an array of pairs of air holes;

[h] Consult the Chapter 11.

diameters of each air hole were 55~60 nm and the center-to-center distance was 71 nm.

Figure 3.32b shows an optical image of the test object, taken by the HLA-implemented microscope. The image are represented with pseudo-color. The left-hand column was relatively light as expected whereas the right-hand column was relatively dark. A hyperlens was located at the position indicated by dashed-line circle. The diameter was same to that observed in Fig. 3.31b. It is evident that a pair of the two tiny objects was observed as two images only under the hyperlens.

Figure 3.32c presents a section profile of the image under the hyperlens in Fig. 3.32b. The measured data are shown with purple dots and the fitted Gaussian curves are shown with red curves. The inset shows the position of the section with red line. The magnified image contains two spots separated by 564.7 nm. Therefore, the magnification was found to be $567.4/71 = 8.0$. The limit of lateral resolution of the microscope is estimated as $L/8.0$. For example, when incident wavelength λ is 405 nm and the NA is 0.9,[i] the limit of lateral resolution is 34 nm from Eq. 3.49. If NA $= 1.2$, $L/8.0 =$ 27 nm. Thus, it has become clear that the limit of lateral resolution reaches a deep subwavelength range less than $\lambda/10$.

One of the features of the HLA-implemented microscope is the low illumination power. From the experimental setup, the net illumination power to object was estimated to be nW/mm^2.

3.6.4 *Application of SMIMs (2): Subwavelength Optical Devices*

SMIMs work as wire-grid polarizers in the long wavelength range. This feature was first pointed out by Hertz (Hertz, 1893). Wire-grid polarizers are now commercially available as handy polarizers working at a wide range. For the near-infrared and visible ranges, wire-grid polarizers are usually made by photolithography and have Al as the metallic component.

---

[i]Objective lens of NA $= 0.9$ can be used without immersion oil and will make the microscope more feasible to use.

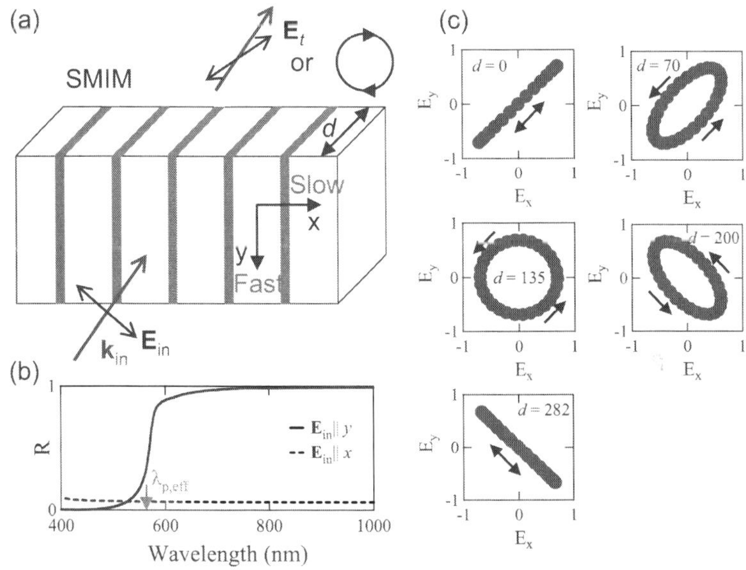

Figure 3.33 Subwavelength waveplate of a SMIM. (a) Schematic and the optical configuration. (b) Linearly polarized R spectra of SMIM of unit cell, (Ag 30 nm/SiO_2 120 nm), under the *x* (dashed line) and *y* (solid line) polarizations. (c) Polarization trajectories projected onto the *xy* plane for various thickness *d*. When $d = 135$ nm, the SMIM serves as a quarter waveplate. When $d = 282$ nm, the SMIM is a half waveplate. Incident wavelength was 460 nm. Adapted with permission from Iwanaga (2008). Copyright 2008, AIP Publishing LLC.

Waveplates of subwavelength thickness Figure 3.33a shows a schematic of SMIM waveplate, which can transform incident linear polarization into circular or 90°-rotated linear polarizations. The waveplates realize extremely thickness in the subwavelength dimension and were numerically designed in a precise manner (Iwanaga, 2008).

Figure 3.33b shows linearly polarized R spectra that were numerically calculated and outlines the working principle of the SMIM waveplates. As an example, the thickness of Ag layer (gray) was set to 30 nm and that of SiO_2 (white) to 120 nm. The *x*-polarized R spectrum (dashed line) keeps low R at the range presented and originates from high-transmission plasmonic waveguide mode addressed in Section 3.1. In contrast, the y-polarized R spectrum (solid line) exhibits high R at the long wavelength range and rapidly decreases below 600 nm. The contrast of polarized T at the long wavelengths is the working principle of the wire-grid polarizers. Here we focus on the shorter wavelength than $\lambda_{\mathrm{p,eff}}$, indicated by an arrow, that is effective plasma wavelength defined by $\varepsilon_y(\lambda_{\mathrm{p,eff}}) = 0$ where ε_y denotes y component of effective permittivity tensor that we can introduce if we regard the SMIM as uniaxial medium. We stress that the SMIM becomes transparent for both x- and y-polarized light. Examining the wavefront and effective wavelength $\lambda/n_{i,\mathrm{eff}}$ ($i = x$ or y) in the SMIM, a relation was found at the transparent range such that

$$0 < n_{y,\mathrm{eff}} < 1 < n_{x,\mathrm{eff}} \tag{3.50}$$

and indicated very strong anisotropy that has not been found in any other conventional media to date. Thus, the x and y axes are interpreted as slow and fast axes, respectively, as indicated in Fig. 3.33a.

Figure 3.33c displays polarization trajectories projected onto the xy plane for various thickness d of the SMIM. At 460 nm, incident polarization was set to linear polarization tilted by 45° from the x and y axes, which is shown in the case of $d = 0$. As the thickness d increases, the polarization of transmitted light becomes elliptic ($d = 70$ nm), circular ($d = 135$ nm), and finally 90°-rotated linear ($d = 282$ nm). It is thus evident that the SMIM works as quarter ($\lambda/4$) and half ($\lambda/2$) waveplates by tuning the thickness. In addition, the SMIMs show high T of 96 and 90% for the $\lambda/4$ and $\lambda/2$ waveplates, respectively. Note that the thickness of SMIM waveplates was in subwavelength range for the incident wavelength of 460 nm. This miniaturization was made possible due to the very strong anisotropy. Other designs of SMIM waveplates for the visible and near-infrared ranges were also found (Iwanaga, 2008).

A few other designs for waveplates have been reported base on more complex unit structures than the SMIM-based design above (for example, Zhao and Alù, 2013; Chen et al., 2015); the experimental results on the complex unit structures have not adequately realized the functions as a waveplate and a polarizer. In comparison with them, the design in Fig. 3.33 keeps advantages in the simplicity and the high T features.

As another elementary optical device, polarizer is well known. The subwavelength polarizer of very high efficiency is described later (see Section 3.8.2).

Circular dichroic (CD) devices of subwavelength thickness Figure 3.34 shows CD devices designed using MM elements; the thickness

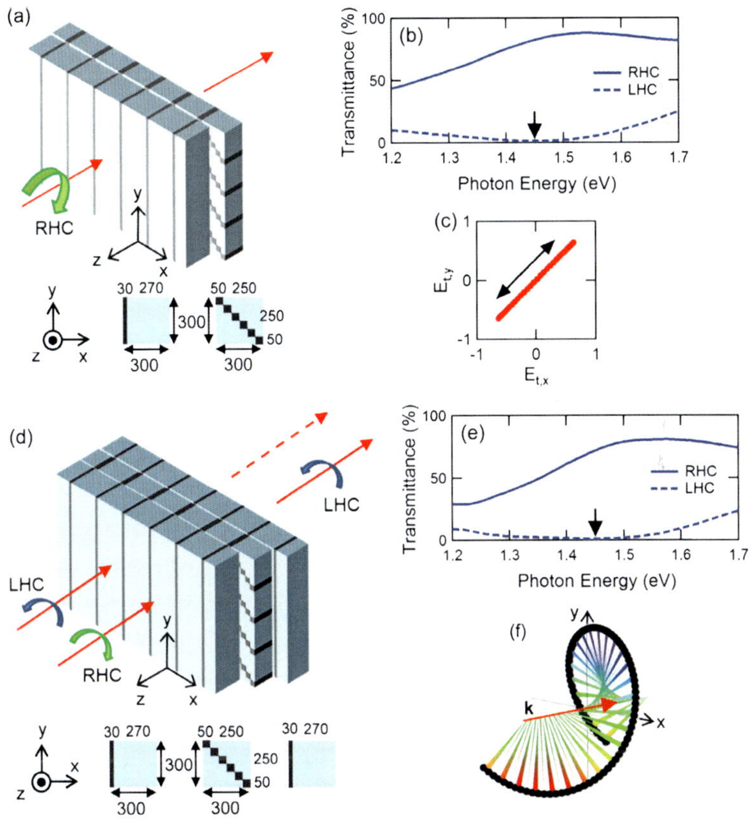

Figure 3.34 Subwavelength CD devices composed of MMs. (a) Two-layer CD device. (b) T spectra of the device (a) for incident RHC (solid line) and LHC (dashed line). (c) E-field vector of the transmitted light; the incidence is RHC and the photon energy is indicated by an arrow in (b). (d) Three-layer CD device. (e) T spectra of the device (d) for incident RHC (solid line) and LHC (dashed line). (f) Snapshot of E-field vector distribution of the transmitted light; the incidence is RHC and the photon energy is indicated by an arrow in (e). Adapted with permission from Iwanaga (2012c).

is subwavelength for the incident light. CD devices generally select optical responses in accordance with incident circular polarizations. We here define right-handed circular (RHC) and left-handed circular (LHC) polarizations such that the polarization vector projected on the *xy* plane forms right-handed and left-handed rotations, respectively, as time *t* goes forward where the temporal phase factor is $\exp(-i\omega t)$.

Figure 3.34a illustrates a two-layer CD device constructed using MM elements. The first layer that incidence sheds on serves as a quarter waveplate, whose working principle was described in Fig. 3.33. The second layer is a polarizer, whose basic property was described in Fig. 3.22. The square unit cells in the *xy* plane are also specified in the units of nm; the periodicity is set to 300 nm. The thickness of the first and second layers are set to 284 and 210 nm, respectively. The constituent materials are set to Ag (gray) and SiO_2 (pale blue).

Figure 3.34b shows numerically calculated T spectra for incidence of RHC and LHC polarizations, shown with blue solid and dashed lines, respectively. T under RHC shows quite large T around 80% whereas T under LHC is suppressed, taking almost 0% at 1.45 eV (855 nm in wavelength), indicated by an arrow. The selection of RHC and LHC light was determined by the design of the two-layer device.

The first layer transforms the incident polarization into linear polarization that satisfies $\mathbf{E}\|(1, 1)$ in the *xy* plane. The second layer works as a polarizer (see Fig. 3.22) and allows transmission for the light of $\mathbf{E}\|(1, 1)$. In total, the two-layer device efficiently works as a CD device. The index σ_{CD} representing circular dichroism is defined by

$$\sigma_{\mathrm{CD}} = \frac{\mathrm{T}_{\mathrm{RHC}} - \mathrm{T}_{\mathrm{LHC}}}{\mathrm{T}_{\mathrm{RHC}} + \mathrm{T}_{\mathrm{LHC}}}. \tag{3.51}$$

where $\mathrm{T}_{\mathrm{RHC}}$ and $\mathrm{T}_{\mathrm{LHC}}$ are transmittance under RHC and LHC incidence, respectively. When the index σ_{CD} satisfies $|\sigma_{\mathrm{CD}}| = 1$, circular dichroism is well realized. This condition is realized at the photon energy indicated by the arrow in Fig. 3.34b. Note that the working wavalength is 855 nm and the total thickness of the two-layer device is 494 nm. Thus, we can call the two-layer device CD device of subwavelength thickness.

Figure 3.34c presents polarization of transmitted light through the two-layer CD device in Fig. 3.34a. Obviously, the polarization is linear, satisfying $\mathbf{E}\|(1, 1)$. Note also that the circular dichroism is realized with high T of approximate 80%.

Figure 3.34d illustrates a three-layer CD device constructed using MM elements. The first and second layers are in common with the two-layer CD device above. The third layer is added and equivalent to the first layer though thickness is adjusted to 255 nm. The total thickness of the three-layer device is 749 nm. This design makes it possible to realize a CD device that transforms the incident circular polarization into the counter circular polarization, described as follows.

Figure 3.34e shows numerically calculated T spectra for incidence of RHC and LHC polarizations, shown with blue solid and dashed lines, respectively. T under RHC incidence exhibits large T values more than 50% at photon energies > 1.4 eV whereas T under LHC incidence is well suppressed around 1.45 eV, indicated by an arrow. This three-layer device also meet the condition of $\sigma_{\mathrm{CD}} = 1$ and a well-defined CD device.

Figure 3.34f displays a snapshot of trajectory of transmitted light polarization. The trajectory form a LHC polarization. It thus turns out that the CD device is able to transform incident circular polarization into the counter circular polarization.

As described in Fig. 3.34, the thickness in both CD device is subwavelength for the incidence. Still, circular dichroism is realized, keeping the T quite high (i.e., several tens %). The efficiency is high and the function is realized in a large degree of freedom; therefore, subwavelength optical devices have potentials to make all the optical devices much miniaturized in the future and such subwavelength optical devices would work as elements in highly functional photonic integrated devices.

The numerical calculations in Fig. 3.34 were carried out with the RCWA+S-matrix method. The permittivity of Ag was taken from literature (Rakić et al., 1998) and that of SiO_2 was set to 2.1316.

3.6.5 *Fishnet MMs*

Fishnet MMs were introduced as optical MMs to exhibiting effective negative refractive index in 2005 (Zhang et al., 2005).

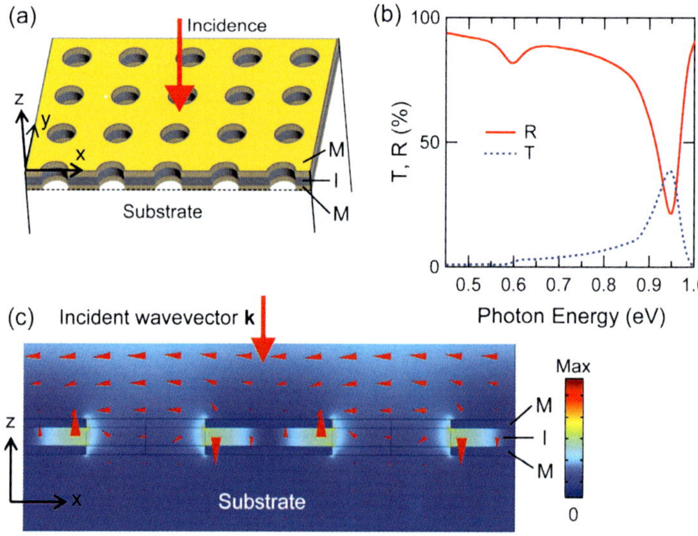

Figure 3.35 Feature summary in fishnet MMs. (a) Illustration of the structure. (b) R (red solid line) and T (blue dotted line) spectra at the normal incidence and *x* polarization. (c) Snapshot of E-field distribution on the resonance at 0.60 eV in the *xz*-section view. The magnitude is plotted with color and the E-field vectors are shown with cones. (a) was adapted from Iwanaga (2012c).

Figure 3.35 summarizes a set of key features of a fishnet MM that is equivalent to the fishnet MM in the first report (Zhang et al., 2005). The fishnet MM attracted great interest as the first example of a MM of effective negative refractive index at optical frequencies (see Section 1.4.3 for the historical aspect).

The structure is a perforated MIM structure on a transparent substrate, as illustrated in Fig. 3.35a. In the original fishnet MM, the metal was Au and the insulator was Al_2O_3, the periodicity was 830 nm, and the diameter of the air holes was typically 400 nm. Calculated R (red solid line) and T (blue dotted line) spectra at the normal incidence of *x* polarization are shown in Fig. 3.35b. In the calculation, the thickness of the MIM layers was set to 30, 60, and 30 nm, respectively, and the substrate was set to SiO_2 of refractive index 1.46. A small R dip at 0.6 eV was found to be related with the effective negative refractive index; the extraction of the index was done using complex reflectivity, complex transmissivity, and the analytic equation such as Eq. 1.66.

Figure 3.35c displays a snapshot of the E field at the condition inducing the R dip at 0.60 eV in Fig. 3.35b. The snapshot is shown in the *xz* section through the centers of the air holes. The color plot denotes the magnitude and the cones indicate the E-field vectors at the points. Obviously, the E field is locally *z*-polarized and prominently enhanced in the I layer. Note that the E_z component is perpendicular to the incident *x* polarization. The E-field distribution indicates that a waveguide mode is excited under the incidence; indeed, the dispersion of the mode was shown to come from the lowest plasmonic waveguide mode in the MIM structure (Iwanaga, 2011, 2012b). The E-field distribution was computed by a finite-element method (FEM) software (COMSOL Multiphysics). Note that the prominent R dip at 0.945 eV comes from a SPP mode reduced (or folded) into the first Brillouin zone in the periodic structure (Iwanaga, 2011).

We here mention the term "fishnet." Although the fishnet MM was originally called double fishnet MM (Zhang et al., 2005), we here call it fishnet MM. This is because more stacking of perforated MIM layers gives rise to similar features, as shown in Fig. 3.36, and there is not any special reason to stress *double*.

Eigenmode analysis The fishnet MM attracted great interest and several theoretical explanations were devoted (Mary et al., 2008; Parsons et al., 2009; García-Meca et al., 2011). As the origin, double resonances of electric and magnetic components were first claimed in accordance with the scenario based on effective permittivity and permeability. However, any linear resonance should be described as a single pole, as described in Section 2.4 (Eq. 2.79).

In further examination on fishnet MMs, the eigenmodes were analyzed (Iwanaga, 2011, 2012b; Cho et al., 2014). Figure 3.36 illustrates a series of results by the eigenmode analysis. Figure 3.36a shows a schematic of fishnet MM standing in air and the optical configuration; the structural parameters such as the periodicity, the diameter of air holes, and thickness of the three layers were set to be same to those in Fig. 3.35.

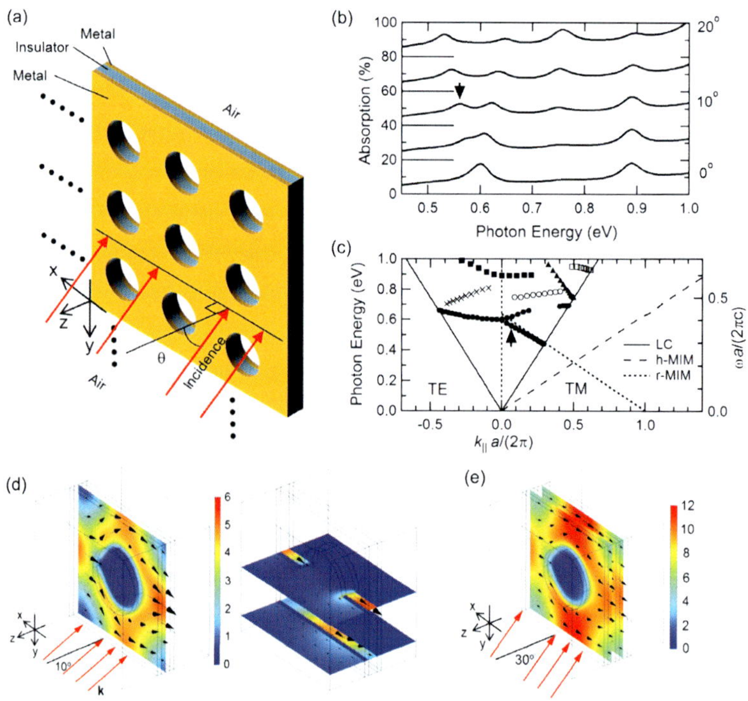

Figure 3.36 A series of eigenmode analysis for a typical fishnet MM. (a) Schematic of a free-standing fishnet MM and the optical configuration. (b) Absorbance spectra dependent on incident angles. Horizontal bars indicate the 0% lines. (c) Dispersion diagram on $(k_{\|}a/(2\pi), \hbar\omega)$ plane. (d) Poynting flow at an oblique incidence. (e) Poynting flow at an oblique incidence in a five-layer stacked fishnet MM. The scales in (d) and (e) are in $\mathrm{W/m^2}$. Adapted from Iwanaga (2012c).

Figure 3.36b shows calculated A spectra dependent on incident angles at p polarization (i.e., $\mathbf{E}_{\mathrm{in}} \| xz$ plane in Fig. 3.36a). When diffraction channels are open, the contributions are also subtracted from the right-hand. The A was defined as follows:

$$\mathrm{A} = \sum_{m,n} 100 - \mathrm{R}_{mn} - \mathrm{T}_{mn} \tag{3.52}$$

in %. Note that Eq. 3.52 is a generalization of Eq. 3.31. The incident angles varied from 0° to 20° in Fig. 3.36b and the A spectra are shown with offset indicated by the horizontal bars.

Figure 3.36c displays dispersion diagram extracted from the A peaks. Since the incident angles were varied in the $0°$–$90°$ range, the data are located inside the light cone (LC) shown with oblique solid lines. The horizontal axis is normalized wavenumber $k_{\parallel}a/(2\pi)$ ($k_{\parallel}$: in-plane wavenumber projected onto the *xy* plane, a: the periodicity) and the vertical axis is photon energy in eV. Two possible TM and TE polarizations were shown.[j] The lowest branch (closed circles) separates into two branches for $k_{\parallel} \neq 0$ at TM (or p) polarization and remains to be a branch at TE (or s) polarization. The lower branch at p polarization was well reproduced by a reduced MIM (r-MIM) waveguide mode shown with a dotted line; the original homogeneous MIM (h-MIM) waveguide mode is plotted with a dashed line, which is described in Section 3.1.1.

In-plane waveguide mode of negative group velocity An interesting suggestion in the dispersion diagram is that the lowest TM mode has negative group velocity at the lower branch. In general, in-plane group velocity $\mathbf{v}_{g,\parallel}$ is expressed as

$$\mathbf{v}_{g,\parallel} = \frac{\partial \omega}{\partial \mathbf{k}_{\parallel}} \tag{3.53}$$

and obviously satisfies with $(\mathbf{v}_{g,\parallel})_x < 0$ at the lower TM branch.

Figure 3.36d presents numerically calculated Poynting flow under the condition that the arrow in Fig. 3.36c indicates: the *xy* section (left) and the *xz* sections (right) are shown. It is evident that the net Poynting flow in the I layer has the opposite direction to the incident in-plane wavevector that is oriented to the $+x$ direction. Input flux was set to 1.45 $\mathrm{W/m^2}$ at the *xy* port of the left edge; as a result, several fold enhancement was observed. The calculation was implemented by the FEM method (see Section 1.5.3).

Figure 3.36e shows Poynting flow in a five-layer stacked fishnet MM, under a similar condition to Fig. 3.36d. It is also evident that in-plane negative group velocity appears in the Poynting flow (Iwanaga, 2012b). The numerical evaluation was also implemented similarly to Fig. 3.36d. About two-fold enhancement appears in comparison

[j]Strictly, it is better to refer to the two polarizations as p and s polarizations, respectively, because the TM and TE polarizations are not exactly separable in 2D periodic systems. The s polarization satisfies with $\mathbf{E}_{\text{in}} \parallel y$ in Fig. 3.36a.

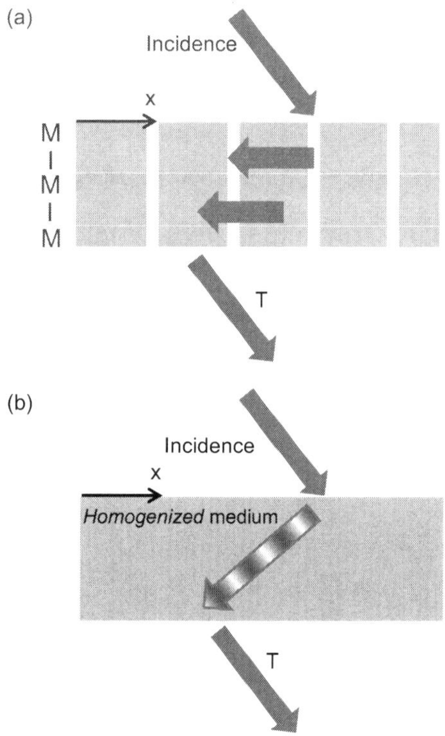

Figure 3.37 Underlying physical mechanism in negative refraction effect in the fishnet MM. (a) Mechanism based on the first-principles analysis. (b) The homogenized view. Reprinted from Iwanaga (2012c).

with Fig. 3.36d. This result originates from the increase in light absorption in the five-layer fishnet MM.

Underlying mechanism in negative refraction effect In addition to the effectively negative refractive index (Zhang et al., 2005), fishnet MMs attracted great interest as media exhibiting negative refraction effect at optical frequencies (Valentine et al., 2008). The experimental demonstration was done by cutting a fishnet MM along an oblique direction by focused ion beam (FIB) and making it a prism shape. Note that negative refraction effect is observed at oblique incidence and is not always linked to negative refractive index because such phenomena can be observed in inhomogeneous light media such as photonic crystals (Kosaka et al., 1998; Notomi, 2000) and nanowire-assembled materials (Yao et al., 2008).

Figure 3.37 schematically illustrates underlying mechanism of the negative refraction effect from the viewpoint of eigenmode analysis. Figure 3.37a draws Poynting flow based on the result in Fig. 3.36e. The backward flow in the fishnet MM is not usual. If one follows the concept of MMs, that is, homogenization of the subwavelength structured media, the fishnet MM is assumed to be a homogeneous medium as illustrated in Fig. 3.37b. Under the assumption, only one possible wavevector is drawn as shown in Fig. 3.37b and corresponds to negative refraction effect inside the fishnet MM. Thus, subwavelength EM dynamics is well clarified without relying on the effective refractive index.

3.7 Plasmonic Resonators with Chirality

MMs also stimulated great interest in chirality because of the diversity by artificial designs. Many attempts were made for the realization. As the most concepts for MMs followed, chiral MMs were also realized at microwaves at the initial stage, and the working wavelengths were moved to THz, and reached an infrared range.

We here see an example of plasmonic resonators with helical structures working at IR wavelengths as a CD device. Figure 3.38 shows a set of SEM images of a Au helix structure that exhibited broadband CD response (Gansel et al., 2009): (a) Vertical section view, which was made using FIB technique, of Au helices (white) and polymer (gray); (b) Oblique view of the Au helix array; (c) Top view. Scale bars (white) indicates 1, 1, and 5 μm in Figs. 3.38a–c, respectively. The periodicity of the array was 2 μm.

The helix structures were fabricated based on a direct laser-writing technique that induced two-photon polymerization of the resist, and made on a glass substrate coated by conductive indium-tin oxide (ITO) layer of 25 nm. After the direct laser writing, plating of Au was conducted using the ITO electrode; this situation was shown in Fig. 3.38a. The polymer was finally removed by plasma etching, and the final structure in Figs. 3.38b and 3.38c was obtained.

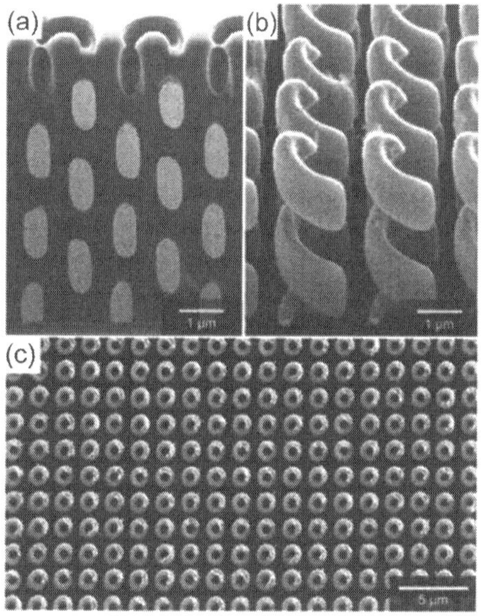

Figure 3.38 A MM exhibiting chiral response (a) SEM image at a vertical section by FIB cutting. (b) SEM image of Au helix array in an oblique view. Scale bars (white) in (a) and (b) indicate 1 μm. (c) Top-view SEM image of the Au helix array. Scale bar (white) indicates 5 μm. From Gansel et al. (2009). Reprinted with permission from AAAS.

The periodic array of left-handed helix showed distinct CD response in transmission spectra. For incidence of left-handed circular polarization, the transmittance T was suppressed less than 10% at a wide IR range of 4.5–8.1 μm. In contrast, for incidence of right-handed circular polarization, the T was more than 80% at the same range. The left-handed circular polarized light is absorbed exciting modes along the helix. The helices in Fig. 3.38 had $360^\circ \times 2.25$ rotations. As the rotations were reduced, the CD range was reduced. Thus, it was suggested that multimode excitations resulted in the broadband CD response.

Comparison with the CD device in Fig. 3.34 The CD response is connected to chirality if homogenized tensor is well defined for the Au helix array. The term chirality presumably makes one imagine helical structures. Consequently, chirality has been so often explored with helical structures.

However, CD response can be designed, in principle, by thinking of the control of polarization and phase. One of the explicit designs was shown in Fig. 3.34.

As for the degree of circular dichroism, $\sigma_{CD} = 1$ in Fig. 3.34 whereas the σ_{CD} of the Au helix array was 0.92 at the maximum. The working wavelength is easily designed in Fig. 3.34 whereas the wavelength tuning is not straightforward in the Au helix array.

The common points are that the two designs successfully exploit the structures in the depth directions and that they include non-resonant wavelength range for high transmission. Thus, apart from the appearance, their working principle has common points.

3.8 Plasmonic Resonators of Stacked Complementary (SC) Structures: Heteroplasmon Hybridized States

3.8.1 *Babinet's Principle*

Babinet's principle states that complementary structures present complementary optical responses. Figure 3.39 shows a typical configuration of Babinet's principle. Disk array of perfect conductor (left) and perforated perfect conductor (right) are complementary in structure. The original Babinet's principle assumes that the

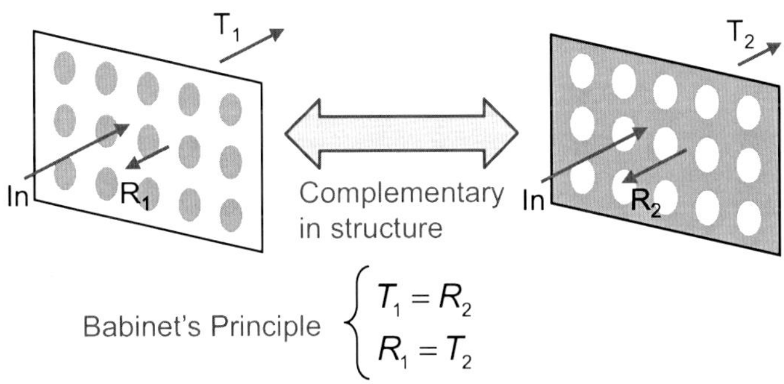

Figure 3.39 Schematics of Babinet's principle, which relates complementary structures to complementary optical responses.

thickness is infinitesimal. In the two structurally complementary systems, optical quantity such as R and T is complementary, satisfying with $T_1 = R_2$ and $R_1 = T_2$ where the indices 1 and 2 denotes the left and right systems, respectively. The relations of R_i and T_i ($i = 1, 2$) represent optical complementary property.

Figure 3.40a shows an illustrative example of Babinet's principle. As a nearly ideal case, we set two configurations: one is perforated

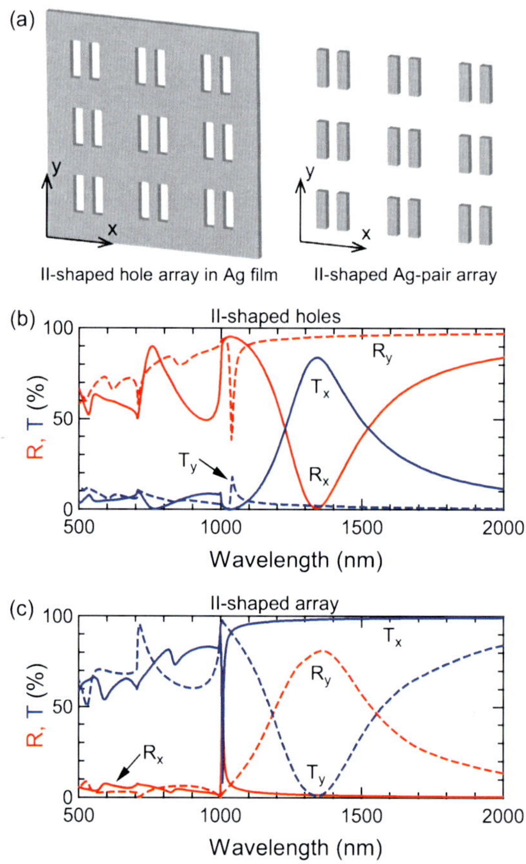

Figure 3.40 Optical responses manifesting Babinet's principle. (a) A set of complementary structures: II-shape perforated metallic film (left) and II-shape array of metal (right). (b) and (c) Numerically calculated optical spectra of Ag film with II-shape holes and of II-shape Ag array, respectively.

metallic film of II-shaped hole array (left), and the other is array of II-shaped metallic pairs (right). The two objects are assumed, for simplicity, to be free-standing in air. We set the periodicity to be 1000 nm and the thickness of the Ag layers to be 50 nm.

Figure 3.40b shows a set of calculated polarized R (red) and T (blue) spectra of the perforated Ag film under the normal incidence. Incident polarizations were set to be x- (solid lines) and y-polarized (dashed lines). In the equivalent optical configuration, a set of polarized R and T spectra of the array of II-shaped Ag pairs is shown in Fig. 3.40c. In the diffraction-free wavelength λ range ($\lambda > 1000$ nm), a good agreement is confirmed such that

$$\mathrm{T}_x \approx \mathrm{R}_y \tag{3.54}$$

$$\mathrm{T}_y \approx \mathrm{R}_x. \tag{3.55}$$

Note that the optical quantities are equal with exchanging the xy polarizations.

Thus, the Babinet's principle stating that complementary structures result in complementary optical responses holds as a good approximation in actual systems including plasmonic resonators. Note that the Babinet's principle is not guaranteed when diffraction channels are open. Indeed, some deviations from Eqs. 3.54 and 3.55 are observed at the wavelength range shorter than 1000 nm in Figs. 3.40b and 3.40c.

The numerical calculation was conducted with the RCWA+S-matrix method. The permittivity of Ag was taken from literature (Rakić et al., 1998) and that of air was set to 1.00054.

3.8.2 *Extremely Anisotropic Local Plasmons*

Strongly anisotropic local plasmons were demonstrated in SC structures (Iwanaga, 2010a,b,c). An experimental result on a SC PlC was referred to in Chapter 1 (Fig. 1.26). The SC PlC consisted of the butterfly-shape array.

Figure 3.41 shows another strongly anisotropic SC PlC of II-shape array, which is schematically illustrated in Fig. 3.41a. The SC PlC is also composed of three layers, that is, top, middle, and bottom layers. The structure can be fabricated through the NIL, as shown in Fig. 1.29. The imprinted resist pattern is seen in Fig. 1.30b.

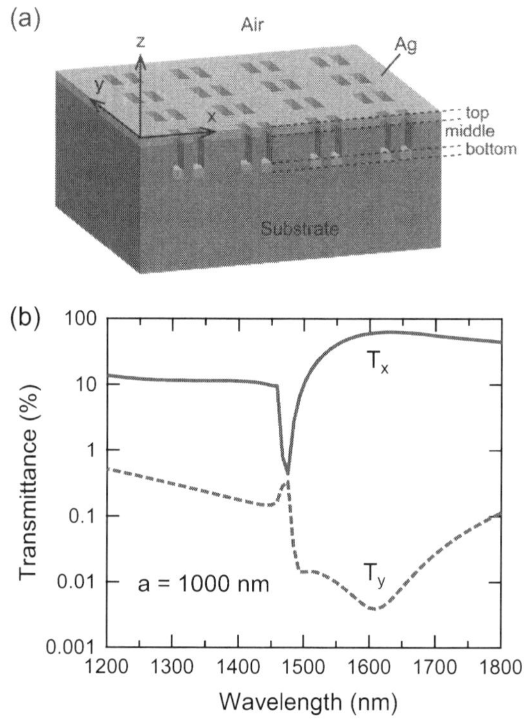

Figure 3.41 (a) Schematics of a SC PlC of II shapes, composed of three stacked layers of top, middle, and bottom. (b) Polarized T spectra calculated numerically. The vertical axis is the logarithmic scale. Solid line: x-polarized T. Dashed line: y-polarized T.

The computed polarized T spectra are shown in Fig. 3.41b. The T is larger for x-polarized light. The significant anisotropy is quantitatively evaluated using extinction ratio, defined by T_x/T_y, exceeds 10^4 around 1600 nm. The periodicity was set to 1000 nm, the II shapes were assumed to be array in a square manner, and the thickness of the top, middle, and bottom layers was 50, 150, and 50 nm, respectively. Thus, the II-shaped SC PlC is an efficient subwavelength polarizer. The metal was set to be Ag and the substrate was assumed to be quartz of refractive index 1.46.

Figure 3.42 shows other structural parameters of the II shapes in the xy plane. Typically, the gap of the two rectangular holes was

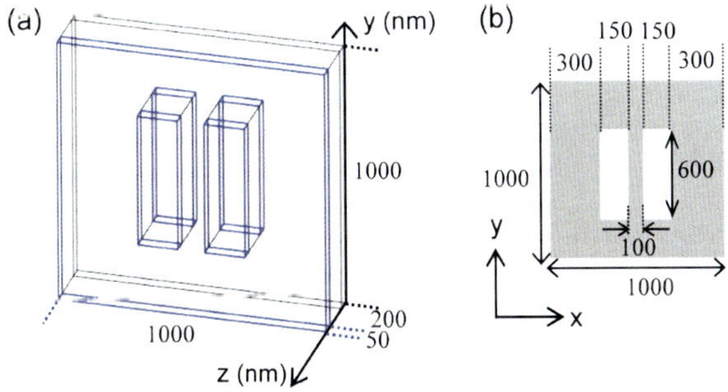

Figure 3.42 Design of unit cell in the SC PIC of II shapes. (a) Unit cell containing the top, middle, and bottom layers. Blue lines indicates the metallic part. (b) Dimensions in the *xy* plane in the units of nm. Adapted with permission from Iwanaga (2010b).

set to 100 nm and the ratio of the rectangular-hole height to the width was 4:1 as shown in Fig. 3.42b, where gray denotes metal. A nanopattern of the II-shaped array produced by UV NIL is seen in Fig. 1.30b.

Figure 3.43 shows resonant EM-field distributions at the high T condition in Fig. 3.41. The unit domain was taken to compute the EM fields; we assigned the *xz* and *yz* boundaries periodic boundary condition. For the *x*-polarized incidence, the E-field and H-field distributions are displayed in Figs. 3.43a and 3.43b, respectively. Both distributions are shown with magnitudes (color plots) and vector fields (cones) at a moment. The E field is mainly enhanced inside the II-shaped air holes whereas the H field is magnified at the backside of the top layer composed of the perforated metal. Incidence was set to be $|\mathbf{E}| = 27$ V/m and $|\mathbf{H}| = 7.3 \times 10^{-2}$ A/m.

Figure 3.43c shows the resonant Poynting vector $\mathbf{S} = \mathbf{E} \times \mathbf{H}^*$; cones indicate the real part of Poynting vectors Re($\mathbf{S}$) and color denotes the flux $|\mathbf{S}|$. The *xz* section is in common with the sections representing the E- and H-fields. The input wave travels from the *xy* port at the left edge. Clearly, Poynting flow goes through the narrow air holes, turns around the II-shaped metals at the bottom layer, and reaches the transmission layer (i.e., substrate). Because the incident wavelength was 1650 nm, these resonant EM flow takes place in the subwavelength dimensions. The computation was implemented using the FEM method (see Section 1.5.3).

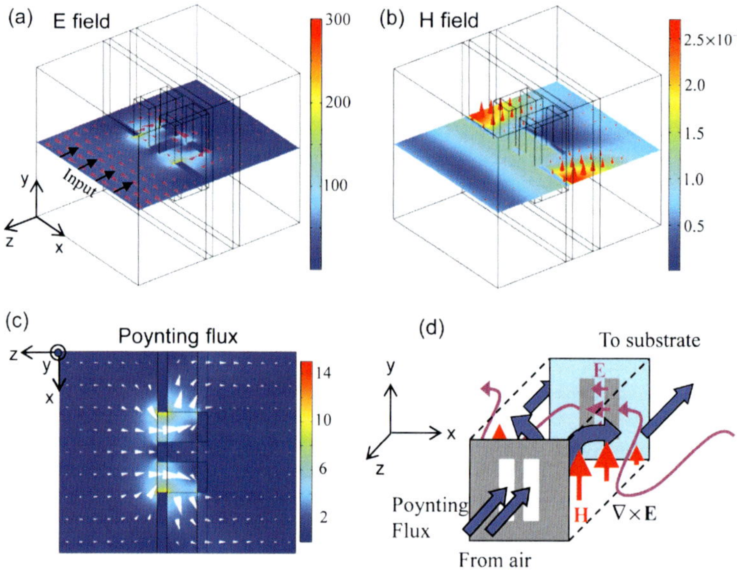

Figure 3.43 Resonant EM-field distributions of a SC PlC of II shapes at the high T condition in Fig. 3.41b. (a) E field. (b) H field. (c) Poynting vector (cone) and the flux (color). (d) A schematic drawing of the resonant local EM fields. Adapted with permission from Iwanaga (2010b).

Figure 3.43d schematically summaries the local resonant EM fields in the subwavelength dimensions. The enhanced E and H fields are represented with light purple and red arrows, respectively. The Poynting flow is indicated by blue arrows. The EM-field distribution represents a resonant mode, which we term local plasmon. Indeed, the $(k_{\|}, \omega)$ dispersion of the resonant mode is independent of $k_{\|}$ where $k_{\|}$ is incident wavenumber projected on the xy plane. Therefore, it is reasonable to call the resonant mode local plasmon. Plasmons are usually characterized only with the E-field component. However, the local plasmon in Fig. 3.43 forms the rotation of E field (i.e., $\nabla \times \mathbf{E}$) and the corresponding H field, so that it is suited to call it rotation-E-type local plasmon. The local plasmon is responsible for the high T for x-polarized light and enables very large extinction ratio exceeding 10^4.

Comparing with the rotation-E-type local plasmon, it is interesting to point out that rotation-H-type local plasmon was also

Accordingly, we call the first mode anti-parallel mode and the third mode parallel mode.

The energy gap between the anti-parallel and parallel modes is described in analogy with a two hetero-atom coupling model. The energy matrix is expressed as

$$\begin{pmatrix} J_{ab} - E & K_a \\ K_b & J_{ba} - E \end{pmatrix}, \tag{3.56}$$

where K_a and K_b are the Coulomb energies of PlasPh modes a and b, respectively. J_{ab} and J_{ba} denote the exchange energy between PlasPh modes a and b. The energy eigenvalues $E_\pm$ were obtained by solving the equation such that the determinant of Eq. 3.56 is equal to 0:

$$E_\pm = \frac{J_{ab} + J_{ba}}{2} \pm \sqrt{K_a K_b - \frac{(J_{ab} + J_{ba})^2}{4}}. \tag{3.57}$$

Energy splitting Δ, defined by $\Delta = E_+ - E_-$, is written as

$$\Delta = \sqrt{4K_a K_b - (J_{ab} + J_{ba})^2}. \tag{3.58}$$

As shown later (Fig. 3.49), the energy gap Δ is approximately 300 meV and large.

Experimental realization of the SC PlasPh crystals Figure 3.47 shows an experimental demonstration of the numerically designed Au SC PlasPh crystal (Choi et al., 2014a). Top-view SEM images are shown in Fig. 3.47a and the inset is the magnified SEM image. A section-view SEM image is shown in Fig. 3.47b. The specimen was fabricated by UV NIL and had 1×1 cm^2 area in the SOI substrate.

Figure 3.47c shows measured R spectra at 24 different spots, which are almost equivalent to each other and indicate the high-uniform nanofabrication. The plane of incidence was the xz plane in Fig. 3.47a, the incident angle was 5°, and the incidence was p-polarized (i.e., $\mathbf{E} \| xz$ plane). The distinct R dips (A and B) correspond to higher-order PlasPh resonances as described later (Fig. 3.50).

Figure 3.47d shows a result on photoluminescence (PL) enhanced by the PlasPh resonances. The red line represents the PL enhancement, defined by the ratio of PL spectrum on the Au PlasPh crystal to the reference PL spectrum measured on Si wafer (inset). The PL enhancement on the PlasPh crystal exceeds 100 times and is quite large. In contrast, the PL enhancement on flat Au film

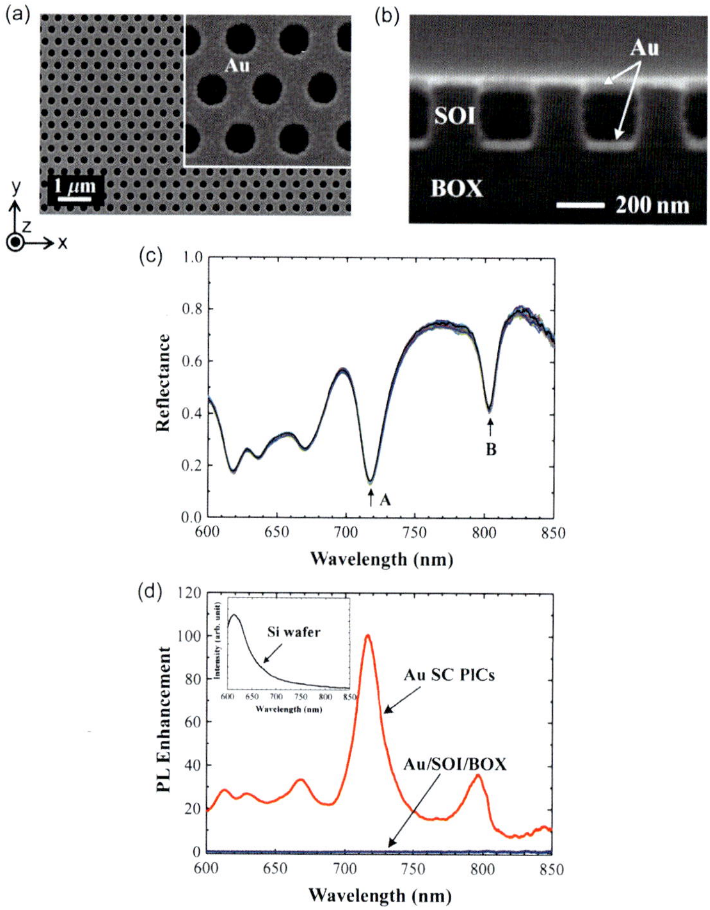

Figure 3.47 (a) Top-view SEM image of a SC PlasPh crystal. Scale bar (white) indicates 1 μm. (b) The section-view SEM image. Scale bar (white) indicates 200 nm. (c) Measured R spectra at 24 different spots on the SC crystal. (d) PL enhancement of dye molecules (rhodamine 590) on the SC PlasPh crystal. The reference was Si wafer. Adapted from Choi et al. (2014a).

deposited on the SOI substrate, shown with blue line, is quite weak and hardly enhanced. Thus, the resonant PL enhancement was explicitly observed, suggesting practical potential to PL sensing of molecules. Further plasmon-enhanced PL (PlasPL) is described in Section 5.3.

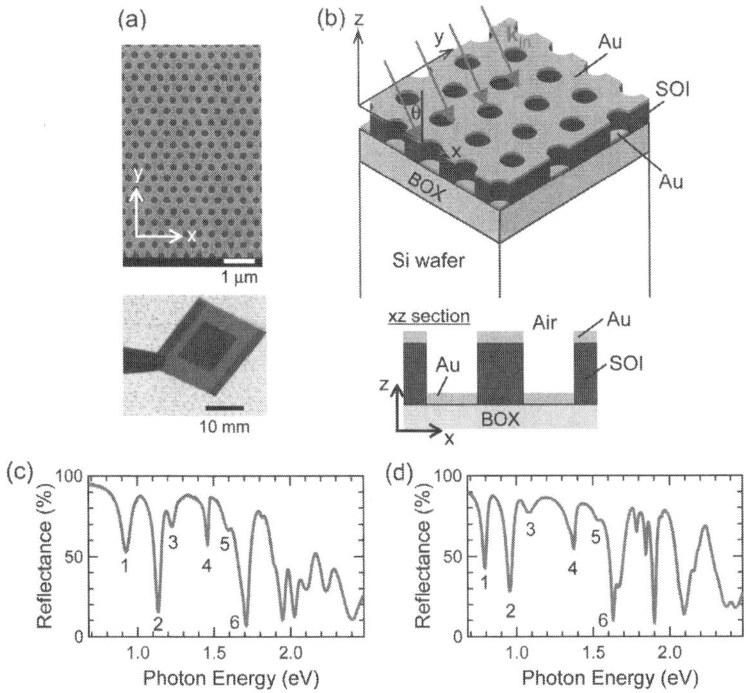

Figure 3.48 SC PlasPh crystal: Experiment versus numerical calculation. (a) The appearance: SEM image (upper) and photograph (lower). (b) Structures and optical configuration. The periodicity was 410.5 nm, the air-hole diameter was 220 nm, and thickness of Au, SOI, and BOX was 42, 250, and 300 nm, respectively. (c) and (d) Measured and calculated R spectra, respectively. Adapted with permission from Iwanaga and Choi (2015). Copyright 2015 American Chemical Society.

A series of PlasPh modes in the SC PlasPh crystal Since the SC PlasPh crystals were first introduced in 2014 (Choi et al., 2014a), the resonant modes were not clarified then. A systematic study in 2015 revealed the resonant modes (Iwanaga and Choi, 2015).

Figure 3.48a shows a top-view SEM image (upper) and a photograph of a SC PlasPh crystal on an SOI substrate. The periodicity of the air-hole array was 410.5 nm and the diameter of the air holes was 220 nm. The SC crystal was fabricated through the UV NIL (Fig. 1.29) on a large area 1×1 cm^2 dimensions in total. In the sample, Au thickness was 42 nm, and the thickness of SOI and BOX layers was 250 and 300 nm, respectively.

Figure 3.48b shows an optical configuration that plane of incidence was set to be the *xz* plane. The upper panel illustrates a 3D view of the PlasPh crystal and the lower panel is the *xz*-section view through the centers of the air holes.

Figure 3.48c shows measured R spectra at incident angle $\theta = 5^\circ$ and p polarization (i.e., $\mathbf{E}_{\text{in}} \| xz$) in the configuration of Fig. 3.48b. The numbers 1 to 6 denote the first to sixth resonant modes. Figure 3.48d is calculated R spectrum corresponding to the measured R spectrum in Fig. 3.48c. It was confirmed that the first to sixth modes are reproduced by the calculation using the RCWA+S-matrix. The permittivity of Au and Si was taken from literature (Rakić et al., 1998; Palik, 1991).

Figures 3.49a–f display numerically calculated snapshots of the resonant EM fields at the first to third modes, respectively. The phase was determined by the incidence that was $E_x \propto \cos(\phi)$ and $\phi = 0^\circ$ at the top edge of PlasPh crystal. As a representative view, we took the EM fields on the *xz* section as shown in Fig. 3.48b. In Fig. 3.49, the E-field (E_x) component is arrayed at the left-hand side and the magnetic (H)-field (H_y) component at the right-hand side. The incidence was set to be normal to the *xy* plane with *x* polarization; accordingly, the E_x and H_y components are presented. The incidence was also set to satisfy $|\mathbf{E}| = 1.0$ and $|\mathbf{H}| = 1.0$. The first to sixth modes are described as follows.

(1) The first mode in Figs. 3.49a and 3.49b represents a hybrid mode of local plasmon at the Au disk located at the bottom of the SC structure and of photonic guided mode in the Si PhC slab. In particular, the resonant EM-field oscillation forms a set of anti-parallel EM field. Therefore we termed the first mode anti-parallel PlasPh mode in Fig. 3.46. Note that the E-field on the *xz* section is included in Fig. 3.46b.
(2) The second mode in Figs. 3.49c and 3.49d represents a PlC-PhC interface mode that is highly localized at the interface of the top and middle layers in the SC crystal. It is interesting that, in terms of the R spectra and EM-field distributions, the second mode is quite similar to the PlasPh mode *a* in Fig. 3.45e, as if PlasPh

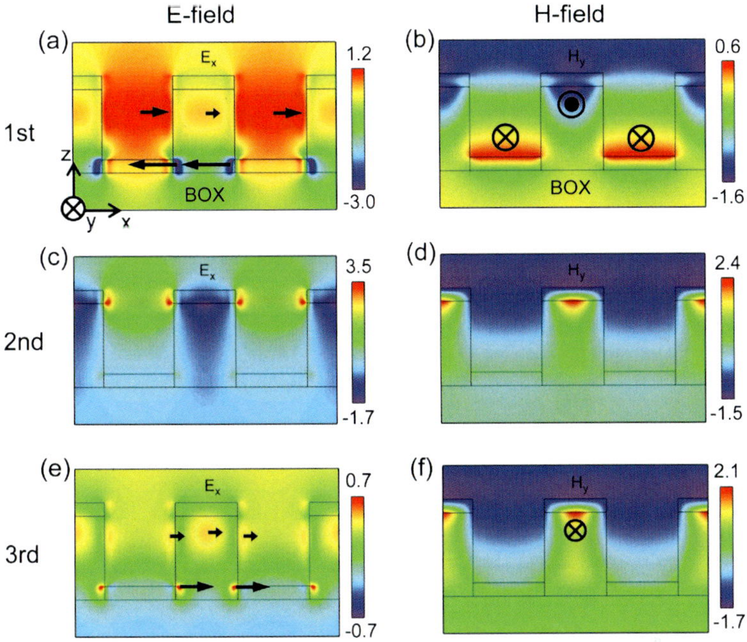

Figure 3.49 PlasPh modes in the SC crystal: the first to third resonances. (a) and (b) E_x and H_y components of the first resonance, respectively. (c) and (d) E_x and H_y of the second resonance, respectively. (e) and (f) E_x and H_y of the third resonance, respectively. These EM fields are snapshots. Adapted with permission from Iwanaga and Choi (2015). Copyright 2015 American Chemical Society.

mode *b* is missing; in fact, the appearance of PlasPh modes *a* and *b* is dependent on the incident configuration, as examined later (Fig. 3.51).

(3) The third mode in Figs. 3.49e and 3.49f represents a hybrid mode of local plasmon at the Au disk located at the bottom of the SC structure and of photonic guided mode in the Si PhC slab. Apart from the first mode, the third mode has a set of parallel EM field; accordingly, we termed the third mode parallel mode in Fig. 3.46.

The energy gap between the first and third modes is 300 meV in the experiment (Fig. 3.48c) and 291 meV in the calculation (Fig. 3.48d), which agree with each other. By using the numerical

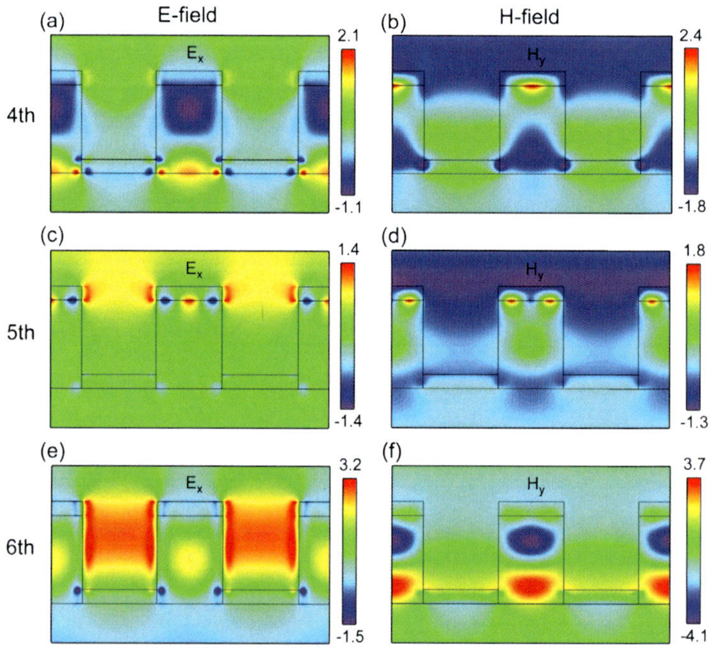

Figure 3.50 Higher PlasPh modes in the SC crystal: the fourth to sixth resonances. The *xz*-section views are displayed similarly to Fig. 3.49. (a) and (b) E_x and H_y components of the fourth resonance, respectively. (c) and (d) E_x and H_y of the fifth resonance, respectively. (e) and (f) E_x and H_y of the sixth resonance, respectively. Adapted with permission from Iwanaga and Choi (2015). Copyright 2015 American Chemical Society.

data, $E_- = 0.788$ and $E_+ = 1.080$ eV, and a relation $K_a \approx K_b$ for Eq. 3.57, we reach $K_a = 0.95$ eV, which is in good agreement with the energy of PlasPh mode a in Fig. 3.45c. Thus, it is to be stressed that the hybridization model self-consistently describes the four energy levels of the anti-parallel, parallel, PlasPh a, and PlasPh b modes.

(4) The fourth mode in Figs. 3.50a and 3.50b is a hybrid mode of a higher local plasmon at the Au disk and of PlasPh a mode. Note that the fourth mode is distinct from the hybrid mode of PlasPh a and b modes.

(5) The fifth mode in Figs. 3.50c and 3.50d is the second-order PlC-PhC interface mode, which corresponds to a shallow dip in the

measured and calculated R spectra in Fig. 3.48. Still, the mode is obviously distinct from other modes.

(6) The sixth mode in Figs. 3.50e and 3.50f is a hybrid mode of a local plasmon at the Au disk and of a higher photonic guided mode in the Si PhC slab.

Overall, all the resonant modes in the SC crystal are hybrid modes of plasmons and photonic guided modes. It is thus proper to call the SC crystal plasmo-photonic crystal. Further properties such as coupling distance of the top and bottom layers and dispersion diagram of the hybrid modes on $(k_{\parallel}, \omega)$ plane ($k_{\parallel}$: wavenumber in the *xy* plane) were described in the literature (Iwanaga and Choi, 2015). The dispersion diagram is suggestive of strong coupling of the two heteroplasmons (i.e., PlasPh modes *a* and *b*).

Direction-selective emittance in SC PlasPh crystals Figure 3.51 describes a SC PlasPh crystal on a transparent substrate of quartz. This figure is shown for clarifying the role of the SOI substrate (i.e., SOI/BOX/Si substrate) in Figs. 3.44–3.50. Figure 3.51a is an illustration at the *xz* section through the centers of air holes, which is similar to Fig. 3.48b. The structural parameters were also similar to Fig. 3.48; the thickness of SOI layer was 250 nm, the diameter of air holes were 220 nm, and the thickness of the Au film was set to 40 nm.

As for the SC PlasPh crystals fabricated on SOI substrates, transmission was negligible, whereas the SC PlasPh crystal on the quartz substrate has practically measurable values of T as shown in Fig. 3.51b (blue and green curves). The R spectra under normal incidence from the air and substrate sides are shown with red solid and dashed curves, respectively. There are several features indicated by the spectra:

(i) The R spectrum under incidence from the air side in Fig. 3.51b is quite similar to the R spectrum in Fig. 3.48d, which means that the PlasPh resonances primarily originate from the nanostructures of the Au and Si slab and that the flat layers of the BOX and Si substrate hardly contribute to the PlasPh resonances.

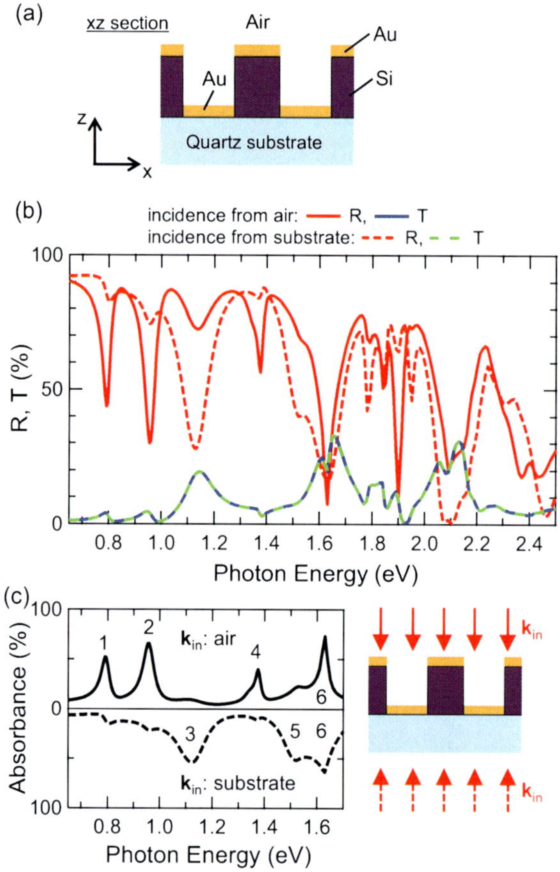

Figure 3.51 SC PlasPh crystal, similar to Fig. 3.48, on a transparent substrate. (a) Schematic: *xz*-section view. (b) R and T spectra by the numerical calculations under incidence from the air (solid lines) and substrate sides (dashed lines). (c) Absorbance for incidence from air (solid line) and substrate sides (dashed line), evaluated based on R and T in (b).

(ii) The T spectra in Fig. 3.51b are exactly same to each other, which results from reciprocity that relates the two optical configurations. As for the reciprocity in optical configurations, a comprehensive review (Potton, 2004) and an intuitive proof (Iwanaga et al., 2007) were reported.

(iii) From the absorbance A defined by Eq. 3.52, it was numerically found that the first to fifth resonances, indicated by 1 to 5 in Fig. 3.51c, are highly dependent on the incident directions; the first, second, and fourth (1, 2, and 4) resonances are effectively excited at the incidence from the air side, while the third and fifth (3 and 5) resonances are efficiently excited at the incidence from the substrate side. This difference reflects the characteristics of each resonance described in Figs. 3.49 and 3.50.

(iv) Higher resonances such as the sixth (6) resonance in Fig. 3.51c show similar amounts of A under the incidence from both air and substrate sides.

(v) Due to Kirchhoff's radiation law (i.e., A = emittance), the A spectra suggest that emittance from the PlasPh crystal is selective for the air or substrate sides. This property is important when considering emittance-related application of the PlasPh crystals to PL enhancement and enhanced Raman scattering.

Figure 3.52a displays a set of top-view SEM images of SC PlasPh crystals, based on SOI substrates and fabricated through the UV NIL with only one mold. The diameters (D) of the air holes were laterally modified by changing the residual resist etching time.

According to the lateral tuning, the resonance of the SC PlasPh crystals was shifted in a systematic way, as shown in Fig. 3.52b. The second resonance shown with mark (2) shifts to shorter wavelengths, as the diameter D becomes larger. The shift of the resonant wavelength was found to be approximately linear with a rate of 2.75 nm per 1 nm diameter change. The structural tuning is useful to set resonant wavelengths to be the designed range for enhanced spectroscopy (Iwanaga et al., 2015).

Underlying Si PhC slab Here we discuss the role of underlying Si PhC slab in the SC PlasPh crystal. Figure 3.53a schematically illustrates a Si PhC slab produce in a SOI substrate, which is structural base of the SC PlasPh crystal in Fig. 3.48, and has 250 nm SOI, 300 nm BOX, and approximately 600 μm Si wafer in the thickness.

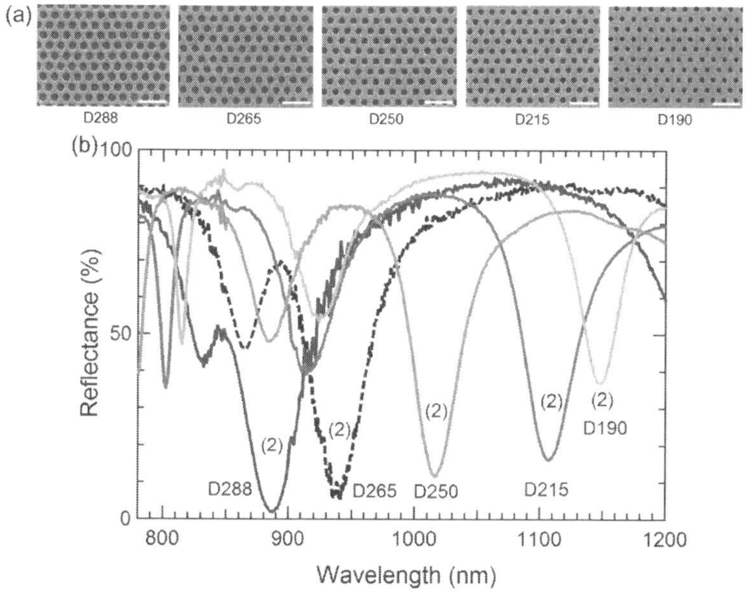

Figure 3.52 Resonance shift of the second PlasPh mode due to structural modification. (a) Top-view SEM images of Au PlasPh crystals with various hole diameters (D) from 288 to 190 nm, which are displayed from left to right, respectively. The series of PlasPh crystals based on SOI substrates was produced using only one mold in the UV NIL procedure. The white scale bars indicate 1 μm. (b) Measured R spectra. The shift of the second (2) mode is displayed in accordance with (a). Adapted from Iwanaga et al. (2015).

Figure 3.53b shows optical density-of-states (DOS) spectra at the Γ point, that is, wavenumber $k_{\|} = 0$ ($k_{\|}$: wavenumber component projected onto the *xy* plane). Red solid line denotes even mode, which is induced from the outside at p polarization. Blue dashed line denotes odd mode, induced at s polarization. Many large DOS modes were found at the frequency range of $a/\lambda > 0.35$ where a is periodicity of the Si PhC slab. The narrow line widths indicate that the modes have high-quality (Q) factors. Arrows indicate large DOS peak frequencies where the corresponding R spectra are studied in the next figure.

Figure 3.53c shows photonic band diagram obtained by the DOS spectra. The vertical dashed arrow indicates the frequency range of present interest. Several bands are located in the range. The lowest

making it clear for a guided mode to be induced in the Si PhC slab. Note that guided modes in Si PhC slabs, which are transverse waves, are usually excited by external plane waves.

Concerning the |**E**| distribution in Fig. 3.56b, it is interesting to point out that the most enhanced |**E**| is observed at the side wall parallel to the *z* axis and includes large E_z component. A similar property to the Ag-disk array in Fig. 3.8c was observed. Thus, longitudinal component of E-field is significantly enhanced also in the Si nanostructures.

3.9 Perfect Absorbers

Perfect absorbers (PAs) have been realized by plasmonic resonators. PAs are considered to be useful because they can collect light energy perfectly. Simultaneously, PAs act as perfect emitter, considered to be useful. The relation of perfect absorption and perfect emittance is often referred to as Kirchhoff's law, and is a consequence from reciprocity at surfaces including very thin artificial structures such as metasurfaces (Greffet and Nieto-Vesperinas, 1998).

PA at GHz Figure 3.57 shows a PA at GHz (Landy et al., 2008): (a) an element is electric resonator made of metallic tip (yellow) in (a); another metallic element is cut wire (yellow) in (b); the design of unit cell for PA is shown in (c), in which the two elements in (a) and (b) are separated by dielectric (pale blue) of 0.72 mm thickness. The typical dimensions were set to $a_1 = 4.2$, $a_2 = 12$, and $G = 0.606$ mm for a working range of 8–12 GHz. The metal was Cu and the thickness of the metallic elements was 17 μm. The design was originally stimulated by the concept of fishnet MMs, described in Section 3.6.5. Incident configuration is also drawn in Fig. 3.57c.

Figure 3.57d shows calculated and measured A spectra. Red line shows a calculated result, exhibiting nearly 100% absorbance and relatively narrow line width at 11.5 GHz. The inset shows the dependence of A on incident angles at the frequency of the A maximum; the perfect absorption is quite sensitive to the incident angles and goes down to 0.3 (i.e., 30%) at 10°. The calculation was done by the FDTD method (see Section 1.5.2). Pale blue and gray

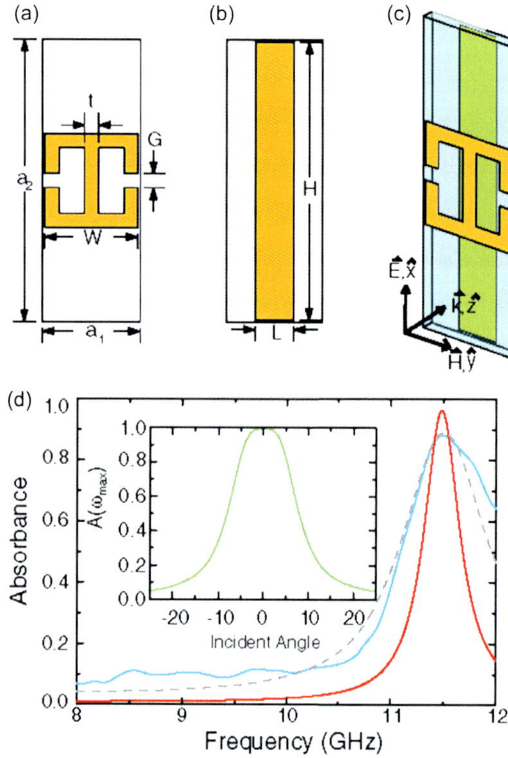

Figure 3.57 PA at GHz. (a) Electric resonator made of metallic tip (yellow). (b) Metallic cut wire (yellow). (c) Design of the unit cell of the PA, in which the two elements are set by separating using dielectric (pale blue). (d) Calculated (red line) and measured (pale blue and gray dashed lines) A spectra shown in the 0–1 range. The inset shows calculated A at 11.5 GHz, dependent on incident angles. Reprinted with permission from Landy et al. (2008). Copyright 2008 by the American Physical Society.

dashed lines represent measured A spectra that reach nearly 90% at the maxima, showing quite a good agreement with the simulation.

The origin of the perfect absorption was also studied by the simulation and found to mainly come from the loss at the dielectric layer between the metallic elements. This property is easily inferred from the resonant EM-field distributions in the MIM-based fishnet MMs we already examined (Figs. 3.35 and 3.36).

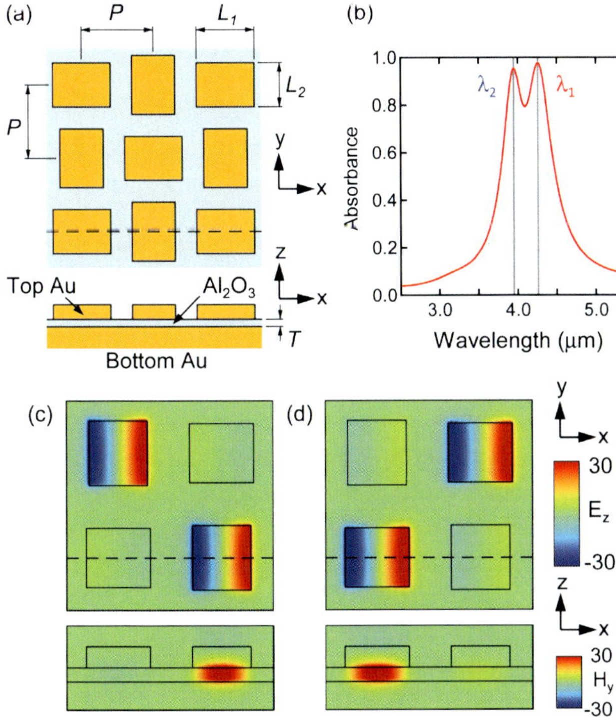

Figure 3.58 PA at a mid-IR range around 4 μm. (a) Design of dual-band PA structure. (b) Numerically evaluated A spectrum. (c) and (d) Resonant EM-field distributions at λ_2 and λ_1, respectively. Adapted with permission from Miyazaki et al. (2014). Copyright 2014, AIP Publishing LLC.

PA at an infrared range Figure 3.58 shows a dual-band PA working at an infrared range around 4 μm. The dual-band PA was specially designed and realized for an in-situ CO_2-sensing IR emitter (Miyazaki et al., 2014).

The design is illustrated in Fig. 3.58a and based on a MIM structure. The top layer comprises submicron Au rectangular structures of $L_1 = 0.93$ μm and $L_2 = 0.85$ μm; the thickness was set to 100 nm. The periodicity (P) was set to 1.5 μm. The I layer was set to be Al_2O_3 of 50 nm thickness. The bottom Au was assumed to be thick enough and flat. We mention that similar designs based on MIM structures were frequently reported because of the feasibility (Hendrickson et al., 2012; Bouchon et al., 2012).

Figure 3.58b shows calculated A spectrum, represented at the 0–1 range. Almost PA was realized at λ_1 (4.26 μm) and λ_2 (3.95 μm) as designed. The two different lengths L_1 and L_2 were tuned to the dual band. The λ_1 light is absorbed by CO_2 molecules in air and the λ_2 light is used as reference. By incorporating the PA into a CO_2 sensing device, low electric-power operation due to only dual-band thermal emission was experimentally demonstrated in comparison with the current commercial devices based on the blackbody radiation (see also Section 5.10).

Figures 3.58c and 3.58d show EM-field distributions at λ_2 and λ_1, respectively, visualizing the underlying mechanism in the PA. The upper panels are snapshots of the E_z component (top view) and the lower the H_y component (section view). Incidence coming from the air side was set to x polarization and the magnitude was normalized. Obviously, the two wavelengths λ_1 and λ_2 correspond to the two lengths L_1 and L_2. Note that similar resonant EM fields are induced for y polarization. Thus, the design for the dual-band PA is independent of polarizations.

PAs at near-infrared and visible ranges Figure 3.59 provides an illustrative example of PA at the near-infrared (NIR) and visible ranges. Figure 3.59a shows a schematic of the PA, which comprises modulated MIM structure on a SiO_2 substrate. The top layer is 1D periodic array along the x axis, composed of metal (Au) of width w_M and air slit of width s. The periodicity is $w_M + s$. The thickness of top, middle, and bottom layers is t_1, t_2, and t_3, respectively. The middle layer was assumed to be Al_2O_3 of permittivity 2.723 and the bottom was Au.

The plane of incidence was set to the xz plane, which is transverse plane to the periodic array of metal. For the 1D periodic structure, incident p polarization, that is, incident E-field vector $\mathbf{E}_{in}$ parallel to the xz plane, excites plasmonic resonances. The incident angle θ is defined as shown in Fig. 3.59a.

Figure 3.59b presents numerically calculated A spectra for incident angles 0° to 30°, shown with black solid, blue dashed, green dot-and-dashed, red dotted curves, respectively. The structural parameters were $w_M = 300$, $s = 100$, $t_1 = t_2 = 40$, and $t_3 = 100$ nm. Incidence was set to p polarization, which satisfied with $\mathbf{E}_{in} \| xz$

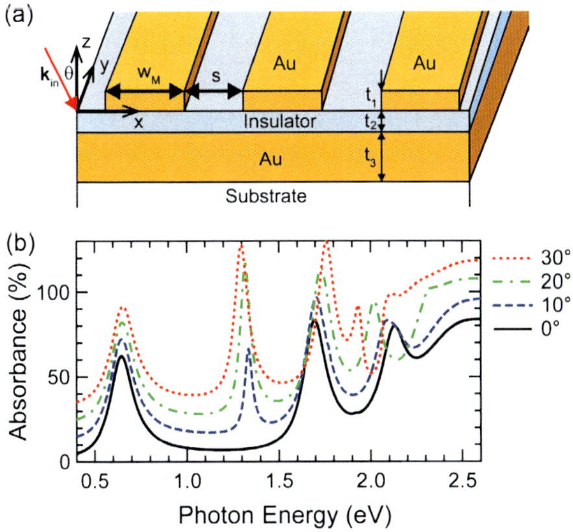

Figure 3.59 PA at the NIR-to-visible range. (a) Illustration of PA of a modulated MIM structure. The top M layer is 1D periodic. (b) Numerically calculated A spectra at incident angles $\theta = 0^\circ$ to 30° (black solid, blue dashed, green dot-and-dashed, and ref dotted lines, respectively). Each A spectrum at $\theta > 0^\circ$ is offset for clarity. Incidence was p-polarized (i.e., $\mathbf{E}_{\text{in}} \| xz$ plane). The structure was set to $w_M = 300$, $s = 100$, $t_1 = t_2 = 40$, and $t_3 = 100$ nm.

plane in Fig. 3.59a. We mention that the A spectra for $\theta > 0^\circ$ are displayed with 10%-step offset for clarity.

The first resonance in energy is almost independent of incident angles, exhibiting a nature of local mode. It is clearly seen that the second resonance appears for $\theta \neq 0^\circ$, suggesting that the mode is an asymmetric mode. The second resonance slightly shifts to lower energies as the θ increases. The third resonance shifts at a small amount to higher energies as the θ increases. Thus, the dispersion of the first to third resonances suggests that they are basically characterized as local modes.

As for light absorption, the first and third resonances are associated with large A at $\theta = 0^\circ$. At $\theta = 30^\circ$, almost 100% absorbance takes place at the second and third resonances for $s = 100$ nm and at third resonance for $s = 200$ nm. The second

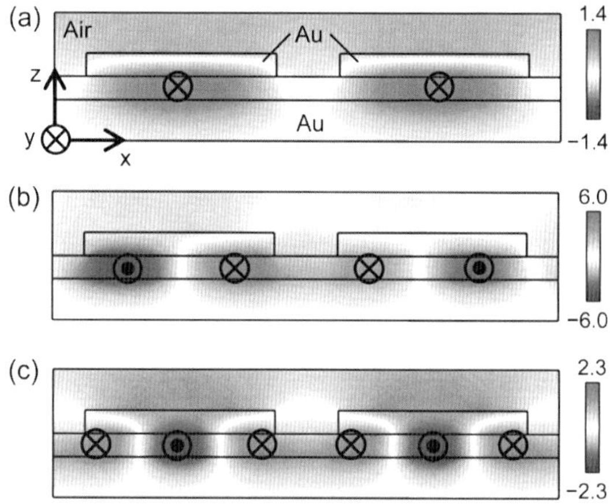

Figure 3.60 (a), (b) and (c) *xz*-section-view snapshots of H_y components at the first, second, and third resonances, respectively. The first resonance was calculated at 0.646 eV and $\theta = 0°$; the second at 1.298 eV and $\theta = 30°$; the third at 1.696 eV and $\theta = 0°$. The incident magnetic field was set to $|\mathbf{H}_{\text{in}}| = 1.0$.

resonances appear at the NIR and visible ranges, and the third resonance at the visible range.

Figures 3.60a–c show snapshots of a H-field component, H_y, at the first, second, and third resonances, respectively. They are displayed in *xz*-section views. The first resonance was evaluated at 0.646 eV and $\theta = 0°$. The second resonance is an asymmetric mode and cannot be excited at $\theta = 0°$, as seen in Fig. 3.60b; accordingly, it was evaluated at 1.298 eV and $\theta = 30°$. The third resonance in Fig. 3.60c was evaluated at 1.696 eV and $\theta = 0°$. The H-field distributions clearly indicate that the first-to-third resonances have the first- to third-order field distributions, respectively. Note that the H_y distributions are useful to visualize the characteristics of 1D periodic systems under p polarization. As for the E-field distribution, the E_z component (not E_x) manifests the localizing feature in the insulator layer. In the insulator layer, the set of (E_z, H_y) describes the characteristics.

We mention the symmetry of the first- to third-order resonances; the first and third resonances display periodic distributions whereas the H_y distribution at the second resonance cannot be periodic along

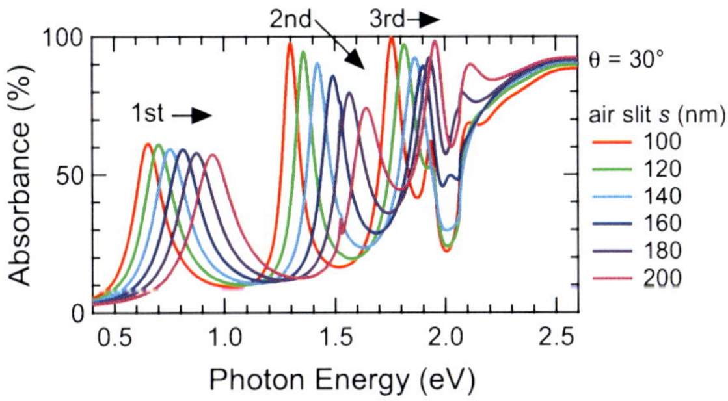

Figure 3.61 The A spectra for various air slit s in Fig. 3.59a. The width of air slit s was varied from 100 to 200 nm at steps of 20 nm. Series of the first to third resonances are indicated by arrows. The directions of the arrows are consistent with the order from the small $s = 100$ to the large $s = 200$ nm. Incident angle θ was fixed at 30°.

the x axis, as easily inferred from Fig. 3.60b. As a result, the second resonance cannot be induced at the normal incidence (i.e., $\theta = 0°$) in Fig. 3.59b. The localized EM fields at the resonances originate mainly from waveguide modes in the MIM structure (see Section 3.1.1) and not from periodic contributions.

Figure 3.61 shows a series of calculated A spectra for various width of air slit s from $s = 100$ to 200 nm at a step of 20 nm, shown with red, green, pale blue, blue, purple, and light purple curves, respectively. Incident angle was fixed at $\theta = 30°$ under p polarization. The periodicity ($w_M + s$) was also fixed at 400 nm. The first to third resonances are noted.

As the slit s increases, the resonance peaks accordingly shift to the higher energies, as indicated with arrows. Obviously, as the s becomes larger (or the metal width w_M smaller), the resonances move to higher energies in continuous ways. Thus, it turns out that the resonant energies are primarily determined by the metal width w_M because the standing-wave-like distributions of the H_y (or E_z) component is proportional to the w_M; the tendency was visualized in Fig. 3.60.

The numerical calculations for Figs. 3.59–3.61 were implemented by the RCWA+S-matrix method (Section 1.5.1). The permittivity of Au was taken from literature (Rakić et al., 1998).

Other structures for PAs In this section, the examples of PAs were based on MIM structures. To obtain better coupling with external EM waves, periodic structures were introduced. Apart from the detailed shapes in the metallic structures, it is understood that the PAs followed a similar strategy using MIM structures. We mention that many plasmonic structures coming from similar idea have been reported.

At the end of this section, let us note other PAs of different structures from the MIM-based ones.

(i) Perfect light absorption was realized in the SC structures (see Section 3.8.3). The resonance was the hybrid type of plasmons and photonic guided resonances, different from the resonances based on MIM structure. Note that the SOI-based SC structures exhibit only reflection and therefore zero R corresponds to perfect absorption by $A = 100 - R$ (%).

(ii) Although there are many other PA structures, another fabricated example is nanoporous Au structures exhibited perfect absorption in an omni-directional way at a range around 750 nm, close to the visible wavelengths. The nanoporus structures were made by Au deposition on latex sphere array and the removal of the sphere array (Teperik et al., 2008).

(iii) Such dielectric sphere arrays were often used to produce a large-area metallic nanostructures in relatively handy ways, in comparison with the EBL procedure (see Section 1.6). As the dielectric spheres in addition to the latex, opal, and polystyrene have been frequently used.

These PAs have been designed or produced for various application such as solar cells, photodetectors, IR emitters, and so on. The IR emitters are addressed in Section 5.10.

3.10 Summary

In this chapter, representative plasmonic resonators have been addressed. Concretely, the followings were described in in-depth ways.

- Plasmonic waveguide modes in the MIM and IMI structures (Section 3.1).
- In periodic structures based on MIM structures, which were fishnet MMs and PAs, the reduced SPPs and plasmonic waveguide modes have been shown to the eigenmodes in the periodic structures (Sections 3.6.5 and 3.9).
- Local plasmons in metallic NPs (Section 3.2).
- Assembled metallic NPs and hybridization of plasmons (Section 3.3).
- Single-layer lattices of simple unit cells (Section 3.4).
- Continuum resonances associated with longitudinal resonant components (Section 3.5).
- Plasmonic resonators of simply stacked structures were addressed; in particular, SMIMs enabling SR imaging and fishnet MMs of effective negative refractive index were described based on the eigenmodes (Section 3.6).
- Chiral plasmonic resonators (Section 3.7).
- Hybrid resonances in SC structures, which included very strong anisotropic local plasmons and PlasPh resonances that are hybridized resonances of plasmons and photonic-guided modes (Section 3.8).

Since some of the plasmonic resonances above have been hardly addressed in textbooks, this book plays a role to shed light on the new aspects of plasmonic resonators.

Making use of macroscopic properties in plasmonic resonators, several applications were also addressed (Sections 5.3.1 and 5.10.1). It is interesting to point out that the applications did not primarily rely on E-field enhancement effect, which is so often stressed in plasmonics. This point of view is also a feature of this book.

References

Adato, R., Yanik, A. A., Amsden, J. J., Kaplan, D. L., Omenetto, F. G., Hong, M. K., Erramilli, S., and Altug, H. (2009). Ultra-sensitive vibrational spectroscopy of protein monolayers with plasmonic nanoantenna arrays, *Proc. Natl. Acad. Sci. U.S.A.* **106**, 19227–19232.

Aouani, H., Šípová, H., Rahmani, M., Navarro-Cia, M., Hegnerová, K., Homola, J., Hong, M., and Maier, S. A. (2013). Ultrasensitive broadband probing of molecular vibrational modes with multifrequency optical antennas, *ACS Nano* **7**, 669–675.

Baida, F. I., Belkhir, A., VanLabeke, D., and Lamrous, O. (2006). Subwavelength metallic coaxial waveguides in the optical range: Role of the plasmonic modes, *Phys. Rev. B* **74**, 205419.

Baida, F. I., and VanLabeke, D. (2003). Three-dimensional structures for enhanced transmission through a metallic film: Annular aperture arrays, *Phys. Rev. B* **67**, 155314.

Bardeen, J., Cooper, L. N., and Schrieffer, J. R. (1957). Theory of superconductivity, *Phys. Rev.* **108**, 1175–1204.

Bennink, R. S., Yoon, Y.-K., Boyd, R. W., and Sipe, J. E. (1999). Accessing the optical nonlinearity of metals with metal–dielectric photonic bandgap structures, *Opt. Lett.* **24**, 1416–1418.

Born, M., and Wolk, E. (1999). *Principles of Optics*, 7th edn. (Cambridge University Press, Cambridge, UK).

Born, P. G. (2013). *Crystallization of Nanoscaled Colloids* (Springer, Berlin).

Bouchon, P., Koechlin, C., Pardo, F., Haïdar, R., and Pelouard, J.-L. (2012). Wideband omnidirectional infrared absorber with a patchwork of plasmonic nanoantennas, *Opt. Lett.* **37**, 1038–1040.

Brown, L. V., Zhao, K., King, N., Sobhani, H., Nordlander, P., and Halas, N. J. (2013). Surface-enhanced infrared absorption using individual cross antennas tailored to chemical moieties, *J. Am. Chem. Soc.* **135**, 3688–3695.

Burgos, S. P., de Waele, R., Polman, A., and Atwater, H. A. (2010). A single-layer wide-angle negative-index metamaterial at visible frequencies, *Nature Mater.* **9**, 407–412.

Carminati, R., Cazé, A., Cao, D., Peragut, F., Krachmalnicoff, V., Pierrat, R., and Wilde, Y. D. (2015). Electromagnetic density of states in complex plasmonic systems, *Surf. Sci. Rep.* **70**, 1–41.

Catrysse, P. B., and Fan, S. (2009). Understanding the dispersion of coaxial plasmonic structures through a connection with the planar metal-insulator-metal geometry, *Appl. Phys. Lett.* **94**, 231111.

Chen, W., Tymchenko, M., Gopalan, P., Ye, X., Wu, Y., Zhang, M., Murray, C. B., Alù, A., and Kagan, C. R. (2015). Large-area nanoimprinted colloidal au nanocrystal-based nanoantennas for ultrathin polarizing plasmonic metasurfaces, *Nano Lett.* **15**, 5254–5260.

Cho, D. J., Wu, W., Wang, F., and Shen, Y. R. (2014). Probing the plasmonic band structure of an optical metamaterial, *Phys. Rev. B* **89**, 035434.

Choi, B., Iwanaga, M., Miyazaki, H. T., Sakoda, K., and Sugimoto, Y. (2014a). Photoluminescence-enhanced plasmonic substrates fabricated by nanoimprint lithography, *J. Micro/Nanolith. MEMS MOEMS* **13**, 023007.

Choi, B., Iwanaga, M., Ochiai, T., Miyazaki, H. T., Sugimoto, Y., and Sakoda, K. (2014b). Subnanomolar fluorescent-molecule sensing by guided resonances on nanoimprinted silicon-on-insulator substrates, *Appl. Phys. Lett.* **105**, 201106.

de Waele, R., Burgos, S. P., Polman, A., and Atwater, H. A. (2009). Plasmon dispersion in coaxial waveguides from single-cavity optical transmission measurements, *Nano Lett.* **9**, 2832–2837.

Drachev, V. P., Podolskiy, V. A., and Kildishev, A. V. (2013). Hyperbolic metamaterials: new physics behind a classical problem, *Opt. Express* **21**, 15048–15064.

Dregely, D., Neubrech, F., Duan, H., Vogelgesang, R., and Giessen, H. (2013). Vibrational near-field mapping of planar and buried three-dimensional plasmonic nanostructures, *Nat. Commun.* **4**, 2237.

Ebbesen, T. W., Lezec, H. J., Ghaemi, H. F., Thio, T., and Wolff, P. A. (1998). Extraordinary optical transmission through sub-wavelength hole arrays, *Nature* **391**, 667–669.

Economou, E. N. (1969). Surface plasmons in thin films, *Phys. Rev.* **182**, 539–554.

Enkrich, C., Wegener, M., Linden, S., Burger, S., Zschiedrich, L., Schmidt, F., Zhou, J. F., Koschny, T., and Soukoulis, C. M. (2005). Magnetic metamaterials at telecommunication and visible frequencies, *Phys. Rev. Lett.* **95**, 203901.

Fan, J. A., Wu, C., Bao, K., Bao, J., Bardhan, R., Halas, N. J., Manoharan, V. N., Nordlander, P., Shvets, G., and Capasso, F. (2010). Self-assembled plasmonic nanoparticle clusters, *Science* **328**, 1135–1138.

Fan, W., Zhang, S., Minhas, B., Malloy, K. J., and Brueck, S. R. J. (2005). Enhanced infrared transmission through subwavelength coaxial metallic arrays, *Phys. Rev. Lett.* **94**, 033902.

Gansel, J. K., Thiel, M., Rill, M. S., Decker, M., Bade, K., Saile, V., von Freymann, G., Linden, S., and Wegener, M. (2009). Gold helix photonic metamaterial as broadband circular polarizer, *Science* **325**, 1513–1515.

García-Meca, C., Hurtado, J., Martí, J., Martínez, A., Dickson, W., and Zayats, A. V. (2011). Low-loss multilayered metamaterial exhibiting a negative index of refraction at visible wavelengths, *Phys. Rev. Lett.* **106**, 067402.

Garcia-Vidal, F. J., Martin-Moreno, L., Ebbesen, T. W., and Kuipers, L. (2010). Light passing through subwavelength apertures, *Rev. Mod. Phys.* **82**, 729–787.

Genet, C., and Ebbesen, T. W. (2007). Light in tiny holes, *Nature* **445**, 39–46.

Giannini, V., Fernández-Domínguez, A. I., Heck, S. C., and Maier, S. A. (2011). Plasmonic nanoantennas: Fundamentals and their use in controlling the radiative properties of nanoemitters, *Chem. Rev.* **111**, 3888–3912.

Greffet, J.-J., and Nieto-Vesperinas, M. (1998). Field theory for generalized bidirectional reflectivity: derivation of helmholtz's reciprocity principle and kirchhoff's law, *J. Opt. Soc. Am. A* **15**, 2735–2744.

Hao, F., Sonnefraud, Y., Dorpe, P. V., Maier, S. A., Halas, N. J., and Nordlander, P. (2008). Symmetry breaking in plasmonic nanocavities: Subradiant lspr sensing and a tunable fano resonance, *Nano Lett.* **8**, 3983–3988.

Hendrickson, J., Guo, J., Zhang, B., Buchwald, W., and Soref, R. (2012). Wideband perfect light absorber at midwave infrared using multiplexed metal structures, *Opt. Lett.* **37**, 371–373.

Hentschel, M., Saliba, M., Vogelgesang, R., Giessen, H., Alivisatos, A. P., and Liu, N. (2010). Transition from isolated to collective modes in plasmonic oligomers, *Nano Lett.* **10**, 2721–2726.

Hertz, H. (1893). *Electric Waves*, translated edn. (Dover, New York).

Hsu, C. W., Zhen, B., Lee, J., Chua, S.-L., Johnson, S. G., Joannopoulos, J. D. and Soljačič, M. (2013). Observation of trapped light within the radiation continuum, *Nature* **499**, 188–191.

Iwanaga, M. (2007). Effective optical constants in stratified metal-dielectric metameterial, *Opt. Lett.* **32**, 1314–1316.

Iwanaga, M. (2008). Ultracompact waveplates: Approach from metamaterials, *Appl. Phys. Lett.* **92**, 153102.

Iwanaga, M. (2010a). Polarization-selective transmission in stacked two-dimensional complementary plasmonic crystal slabs, *Appl. Phys. Lett.* **96**, 083106.

Iwanaga, M. (2010b). Subwavelength electromagnetic dynamics in stacked complementary plasmonic crystal slabs, *Opt. Express* **18**, 15389–15398.

Iwanaga, M. (2010c). Electromagnetic eigenmodes in a stacked complementary plasmonic crystal slab, *Phys. Rev. B* **82**, 155402.

Iwanaga, M. (2011). In-plane plasmonic modes of negative group velocity in perforated waveguides, *Opt. Lett.* **36**, 2504–2506.

Iwanaga, M. (2012a). *Collective Plasmonic States Emerged in Metallic Nanorod Array and Their Application* in *Nanorods* (InTech, Rijeka, Croatia), 75–92.

Iwanaga, M. (2012b). First-principle analysis for electromagnetic eigen modes in an optical metamaterial slab, *Prog. Electromagn. Res.* **132**, 129–148.

Iwanaga, M. (2012c). Photonic metamaterials: a new class of materials for manipulating light waves, *Sci. Technol. Adv. Mater.* **13**, 053002.

Iwanaga, M. (2014). Hyperlens-array-implemented optical microscopy, *Appl. Phys. Lett.* **105**, 053112.

Iwanaga, M. (2015). Toward super-resolution imaging at green wavelengths employing stratified metal-insulator metamaterials, *Photonics* **2**, 468–482.

Iwanaga, M., and Choi, B. (2015). Heteroplasmon hybridization in stacked complementary plasmo-photonic crystals, *Nano Lett.* **15**, 1904–1910.

Iwanaga, M., Choi, B., Miyazaki, H. T., Sugimoto, Y., and Sakoda, K. (2015). Large-area resonance-tuned metasurfaces for on-demand enhanced spectroscopy, *J. Nanomater.* **2015**, 507656.

Iwanaga, M., Ikeda, N., and Sugimoto, Y. (2012). Enhancement of local electromagnetic fields in plasmonic crystals of coaxial metallic nanostructures, *Phys. Rev. B* **85**, 045427.

Iwanaga, M., Vengurlekar, A. S., Hatano, T., and Ishihara, T. (2007). Reciprocal transmittances and reflectances: An elementary proof, *Am. J. Phys.* **75**, 899–902.

Jacob, Z., Alekseyev, L. V., and Narimanov, E. (2006). Optical hyperlens: Far-field imaging beyond the diffraction limit, *Opt. Express* **14**, 8247–8256.

Jin, R., Cao, Y., Mirkin, C. A., Kelly, K. L., Schatz, G. C., and Zheng, J. G. (2001). Photoinduced conversion of silver nanospheres to nanoprisms, *Science* **294**, 1901–1903.

Kilmov, V. (2014). *Nanoplasmonics* (Pan Stanford Publishing, Singapore).

Kinkhabwala, A., Yu, Z., Fan, S., Avlasevich, Y., Müllen, K., and Moerner, W. E. (2009). Large single-molecule fluorescence enhancements produced by a bowtie nanoantenna, *Nature Photon.* **3**, 654–657.

Kosaka, H., Kawashima, T., Tomita, A., Notomi, M., Tamamura, T., Sato, T., and Kawakami, S. (1998). Superprism phenomena in photonic crystals, *Phys. Rev. B* **58**, R10096–R10099.

Kuwata, H., Tamaru, H., Esumi, K., and Miyano, K. (2003). Resonant light scattering from metal nanoparticles: Practical analysis beyond rayleigh approximation, *Appl. Phys. Lett.* **83**, 4625–4627.

Landy, N. I., Sajuyigbe, S., Mock, J. J., Smith, D. R., and Padilla, W. J. (2008). Perfect metamaterial absorber, *Phys. Rev. Lett.* **100**, 207402.

Li, G., Cherqui, C., Wu, Y., Bigelow, N. W., Simmons, P. D., Rack, P. D., Masiello, D. J., and Camden, J. P. (2015). Examining substrate-induced plasmon mode splitting and localization in truncated silver nanospheres with electron energy loss spectroscopy, *J. Phys. Chem. Lett.* **6**, 2569–2576.

Linden, S., Enkrich, C., Wegener, M., Zhou, J., Koschny, T., and Soukoulis, C. M. (2004). Magnetic response of metamaterials at 100 terahertz, *Science* **306**, 1351–1353.

Liu, Z., Lee, H., Xiong, Y., Sun, C., and Zhang, X. (2007). Far-field optical hyperlens magnifying sub-diffraction-limited objects, *Science* **315**, 1686.

Mary, A., Rodrigo, S. G., Garcia-Vidal, F. J., and Martin-Moreno, L. (2008). Theory of negative-refractive-index response of double-fishnet structures, *Phys. Rev. Lett.* **101**, 103902.

Maxwell-Garnett, J. C. (1904). Colours in metal glasses and in metallic films, *Philos. Trans. Royal Soc. London A* **203**, 385–420.

Maxwell-Garnett, J. C. (1906). Colours in metal glasses, in metallic films, and in metallic solutions. II, *Philos. Trans. Royal Soc. London A* **205**, 237–288.

Miyazaki, H. T., Kasaya, T., Iwanaga, M., Choi, B., Sugimoto, Y., and Sakoda, K. (2014). Dual-band infrared metasurface thermal emitter for CO_2 sensing, *Appl. Phys. Lett.* **105**, 121107.

Mock, J. J., Barbic, M., Smith, D. R., Schultz, D. A., and Schultz, S. (2002). Shape effects in plasmon resonance of individual colloidal silver nanoparticles, *J. Chem. Phys.* **116**, 6755–6759.

Murray, A., and Barnes, W. L. (2007). Plasmonic materials, *Adv. Mater.* **19**, 3771–3782.

Neubrech, F., Pucci, A., Cornelius, T. W., Karim, S., García-Etxarri, A. and Aizpurua, J. (2008). Resonant plasmonic and vibrational coupling in a tailored nanoantenna for infrared detection, *Phys. Rev. Lett.* **101**, 157403.

Notomi, M. (2000). Theory of light propagation in strongly modulated crystals: refraction-like behavior in the vicinity of the photonic band gap, *Phys. Rev. B* **62**, 10696–10705.

Novotny, L., and Hecht, B. (2012). *Principles of Nano-Optics*, 2nd edn. (Cambridge University Press, New York).

O'Donnell, K. A., and Torre, R. (2005). Characterization of the second-harmonic response of a silver-air interface, *New J. Phys.* **7**, 154.

Ohtaka, K., Inoue, J., and Yamaguti, S. (2004). Derivation of the density of states of leaky photonic bands, *Phys. Rev. B* **70**, 035109.

Palik, E. D. (1991). *Handbook of Optical Constants of Solids II* (Academic, San Diego, USA).

Parsons, J., Hendry, E., Sambles, J. R., and Barnes, W. L. (2009). Localized surface-plasmon resonances and negative refractive index in nanostructured electromagnetic metamaterials, *Phys. Rev. B* **80**, p. 245117.

Pendry, J. B. (2000). Negative refraction makes a perfect lens, *Phys. Rev. Lett.* **85**, 3966–3969.

Poddubny, A., Iorsh, I., Belov, P., and Kivshar, Y. (2013). Hyperbolic metamaterials, *Nature Photon.* **7**, 958–967.

Potton, R. J. (2004). Reciprocity in optics, *Rep. Prog. Phys.* **67**, 717–754.

Poujet, Y., Salvi, J., and Baida, F. I. (2007). 90% extraordinary optical transmission in the visible range through annular aperture metallic arrays, *Opt. Lett.* **32**, 2942–2944.

Prodan, E., Radloff, C., Halas, N. J., and Nordlander, P. (2003). A hybridization model for the plasmon response of complex nanostructures, *Science* **302**, 419–422.

Rakić, A. D., Djurušić, A. B., Elazar, J. M., and Majewski, M. L. (1998). Optical properties of metallic films for vertical-cavity optoelectronic devices, *Appl. Opt.* **37**, 5271–5283.

Ramakrishna, S. A., and Pendry, J. B. (2003). Removal of absorption and increase in resolution in a near-field lens via optical gain, *Phys. Rev. B* **67**, 201101(R).

Rho, J., Ye, Z., Xiong, Y., Yin, X., Liu, Z., Choi, H., Bartal, G., and Zhang, X. (2010). Spherical hyperlens for two-dimensional sub-diffractional imaging at visible frequencies, *Nat. Commun.* **1**, 143.

Salandrino, A., and Engheta, N. (2006). Far-field subdiffraction optical microscopy using metamaterial crystals: Theory and simulations, *Phys. Rev. B* **74**, 075103.

Scalora, M., Bloemer, M. J., Pethel, A. S., Dowling, J. P., Bowden, C. M., and Manka, A. S. (1998). Transparent, metallo-dielectric, one-dimensional, photonic band-gap structures, *J. Appl. Phys.* **83**, 2377–2383.

Scalora, M., Vincenti, M. A., de Ceglia, D., Roppo, V., Centini, M., Akozbek, N., and Bloemer, M. J. (2010). Second- and third-harmonic generation in metal-based structures, *Phys. Rev. A* **82**, 043828.

Selig, O., Siffels, R., and Rezus, Y. L. A. (2015). Ultrasensitive ultrafast vibrational spectroscopy employing the near field of gold nanoantennas, *Phys. Rev. Lett.* **114**, 233004.

Smolyaninov, I. I., Hung, Y.-J., and Davis, C. C. (2007). Magnifying superlens in the visible frequency range, *Science* **315**, 1699–1701.

Swihart, J. C. (1961). Field solution for a thin-film superconducting strip transmission line, *J. Appl. Phys.* **32**, 461–469.

Teperik, T. V., de Abajo, F. J. G., Borisov, A. G., Abdelsalam, M., Bartlett, P. N., Sugawara, Y., and Baumberg, J. J. (2008). Omnidirectional absorption in nanostructured metal surfaces, *Nature Photon.* **2**, 299–301.

Valentine, J., Zhang, S., Zentgraf, T., Ulin-Avila, E., Genov, D. A., Bartal, G., and Zhang, X. (2008). Three-dimensional optical metamaterial with a negative refractive index, *Nature* **455**, 376–379.

Verhagen, E., de Waele, R., Kuipers, L., and Polman, A. (2010). Three-dimensional negative index of refraction at optical frequencies by coupling plasmonic waveguides, *Phys. Rev. Lett.* **105**, 223901.

Xu, T., Agrawal, A., Abashin, M., Chau, K. J., and Lezec, H. J. (2013). All-angle negative refraction and active flat lensing of ultraviolet light, *Nature* **497**, 470–474.

Yao, J., Liu, Z., Liu, Y., Wang, Y., Sun, C., Bartal, G., Stacy, A. M., and Zhang, X. (2008). Optical negative refraction in bulk metamaterials of nanowires, *Science* **321**, 930.

Yeh, P., Yariv, A., and Hong, C.-S. (1977). Electromagnetic propagation in periodic stratified media. I. General theory, *J. Opt. Soc. Am.* **67**, 423–438.

Yen, T. J., Padilla, W. J., Fang, N., Vier, D. C., Smith, D. R., Pendry, J. B., Basov, D. N., and Zhang, X. (2004). Terahertz magnetic response from artificial materials, *Science* **303**, 1494–1496.

Zentgraf, T., Dorfmüller, J., Rockstuhl, C., Etrich, C., Vogelgesang, R., Kern, K., Pertsch, T., Lederer, F., and Giessen, H. (2008). Amplitude- and phase-resolved optical near fields of split-ring-resonator-based metamaterials, *Opt. Lett.* **33**, 848–850.

Zhang, S., Fan, W., Panoiu, N. C., Malloy, K. J., Osgood, R. M., and Brueck, S. R. J. (2005). Experimental demonstration of near-infrared negative-index metamaterials, *Phys. Rev. Lett.* **95**, 137404.

Zhao, Y., and Alù, A. (2013). Tailoring the dispersion of plasmonic nanorods to realize broadband optical meta-waveplates, *Nano Lett.* **13**, 1086–1091.

Chapter 4

Nonlocality on Plasmonic Resonances

Plasmon is one of collective excitations described in the classical regime. However, the constituent electrons follow, in principles, the quantum mechanical regime. Therefore, the limit of the classical description already attracted great interest at the early stage of plasmon studies in 1980s. After the extensive development of plasmonics and metamaterials in 2000s, it was an issue to determine where the boundary should be properly set for plasmonic resonators. In this chapter, we survey the classical limit that has been clarified based on experimental results.

Figure 4.1 schematically illustrates boundary of the quantum and classical regimes in plasmonic resonators. As an indicator to evaluate the boundary, we here introduce characteristic dimension that represents spatial dimension of metallic nanostructures inducing plasmonic resonances. Let us describe how the boundary was finally drawn at approximate 1 nm.

Plasmonic Resonators: Fundamentals, Advances, and Applications
Masanobu Iwanaga

ISBN 978-981-4745-06-2 (Hardcover), 978-981-4745-07-9 (eBook)
www.panstanford.com

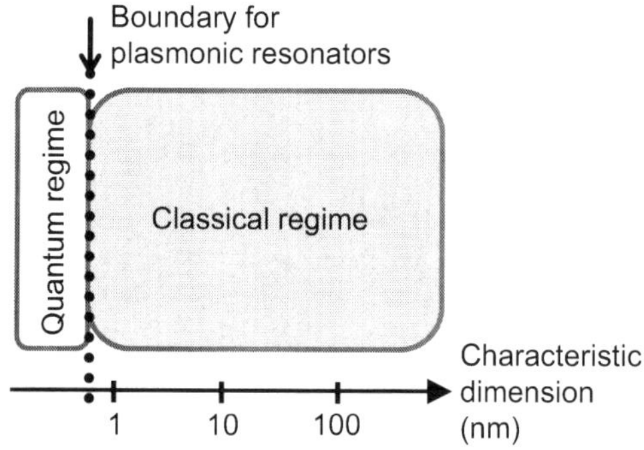

Figure 4.1 Boundary of the quantum and classical regimes in plasmonic resonators. Characteristic dimension indicates the size of metallic nanostructure inducing each plasmonic resonance.

4.1 Nonlocal Responses in Far-Field Spectra

Nonlocal responses of crystalline metals were explored in metal optics and were already observed in 1980s (Forstmann and Gerhardts, 1986). In this section, the experimental evidences and the corresponding theoretical analyses are described. Besides, the implications are discussed.

Modified electron density at metal surface To detect prominent change of optical signals, electroreflectance (ER) measurement was conducted in 1980s. Figure 4.2 illustrates the typical configuration and concept. Let us consider a situation that bulk metal of single crystal is set in air, incident light of wavevector $\mathbf{k}_{\text{in}}$ is illuminated on a side, and the other side is connected to a direct-current voltage source. By applying the voltage, the electron density at the surface of the bulk metal is changed. The other electrode of the voltage source was set in 0.5 mol/l $NaClO_4$. Phenomenologically, the modulated thin layer near the surface is assumed to have permittivity ε_2, which is different from permittivity in the bulk metal ε_3 because of the voltage-modified electron density.

Measured physical quantity in the ER experiment was relative reflectance change $\Delta R/R$, which is defined by

$$\frac{\Delta R}{R} = \frac{R_V - R_{\text{pzc}}}{R_{\text{pzc}}} \tag{4.1}$$

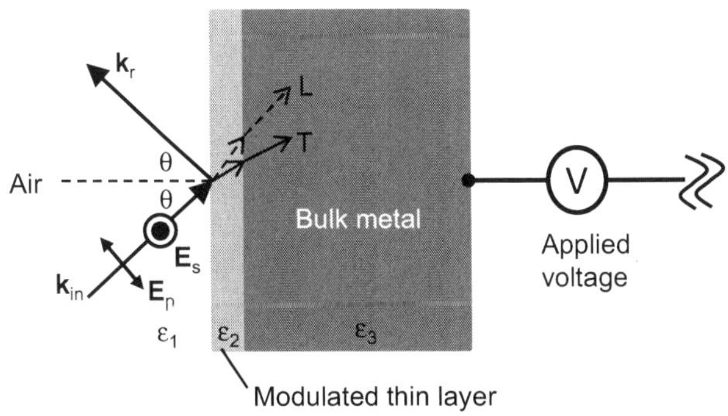

Figure 4.2 Schematic of a typical electroreflectance (ER) configuration. Modulated thin film layer was assumed as a consequence of the applying voltage.

where R_{pzc} denotes reflectance near the potential zero charge and R_V reflectance at bias potential V, which was varied from -1.1 to $+0.1$ V. The voltage for the potential zero charge was -0.7 V for the other electrode in the $NaClO_4$ solution.

Transverse and longitudinal components of relative permittivity of metal When one approximates the electrons in metals as free electrons, Drude permittivity is derived for bulk metals as described in Section 1.2.2. However, when the metals have flat boundary, the relative permittivity has two components, i.e., transverse and longitudinal components.

The transverse component, which is here written as $\varepsilon_T(\omega)$, is same to the Drude permittivity in Eq. 1.24 for ideal Drude metal. Because we now consider that bulk metal is Ag single crystal, the permittivity is approximated using Lorentz-Drude model[a] such that

$$\varepsilon_T(\omega) = \varepsilon_{\text{other}}(\omega) - \frac{\omega_p^2}{\omega(\omega + i\gamma)} \tag{4.2}$$

where $\varepsilon_{\text{other}}$ contains other terms in the Lorentz-Drude permittivity except for the ω-dependent term in the Drude permittivity.

[a]See Section 1.2.3.

On the other hand, the longitudinal component $\varepsilon_L(\omega, \mathbf{k})$ is expressed as

$$\varepsilon_L(\omega, \mathbf{k}) = \varepsilon_{\text{other}}(\omega) - \frac{\omega_p^2}{\omega(\omega + i\gamma) - \beta k_1^2}, \tag{4.3}$$

assuming free electron gas. This **k**-dependent permittivity indicates nonlocal response of metal. In comparison with Eq. 4.2, the difference lies in the contribution of the k_1^2 factor. Here the factor β is defined by

$$\beta = \frac{3}{5} v_F^2$$

where v_F is Fermi velocity of the electron. The k_1 is defined to satisfy $\varepsilon_L(\omega, \mathbf{k}_1) = 0$ and $k_1 = |\mathbf{k}_1|$; therefore, the relation to determine k_1 is obtained as follows.

$$k_1^2 = \frac{\omega(\omega + i\gamma) - \omega_p^2/\varepsilon_{\text{other}}}{\beta}. \tag{4.4}$$

In the bias-modulated thin layer of ε_2, the EM-wave components tangential to the surface is described using ε_T, just as the Drude metal, whereas the component perpendicular to the surface is described using ε_L. Thus, the thin layer serves as uniaxial material[b] and allows birefringence, which is conceptually illustrated in Fig. 4.2.

It is interesting to point out that, even in the configuration composed of only one flat Ag–air interface, the longitudinal component ε_L was found to be included in the following analysis. In Sections 2.2 and 2.4, we noted the longitudinal component in the configuration interaction and the nonlocal theory, respectively, though it is often ignored in comparison with the transverse component.

Results and analysis on ER Figures 4.3a and 4.3b show measured and calculated ER spectra, i.e., the modulated reflectance change $\Delta R/R$, originally reported in 1980 (Kötz et al., 1980).

The optical configuration of the ER measurement was drawn in Fig. 4.2. Incident angle θ was set to 45°. The incidence was set to p and s polarizations. The Ag surface was the (100) plane. ER spectra

[b]The permittivity written as ε_2 in Eq. 4.2 turns out to have a uniaxial tensor containing ε_T and ε_L.

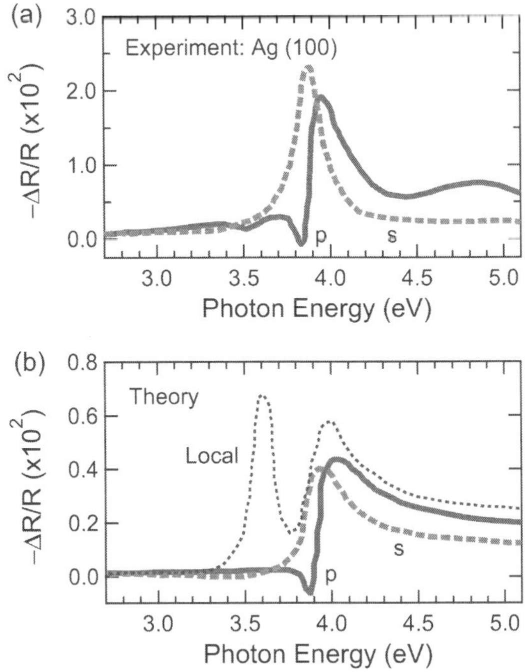

Figure 4.3 ER spectra for testing nonlocal response of Ag single crystals. (a) Experimental results. Incident angle was 45°. (b) Theoretical results. Solid lines designate ER spectra for p polarization. Dashed lines present ER spectra for s polarization. Dotted line in (b) shows calculated p-polarized ER spectrum based on the local theory. The data were taken from Kötz et al. (1980).

at p and s polarizations are shown with solid and dashed lines, respectively. Major signals appear at a range of 3.5–4.5 eV. The bias voltage was −0.9 V. The effective plasma frequency ω_p of Ag, which is simply determined by $\varepsilon_{\text{Ag}}(\omega_p) = 0$, is located at 3.8 eV. As noted in Section 1.2.3, the frequency ω_p corresponds to the interband *d*–*d* transition. The resonance was found to contribute to the prominent signals of ΔR/R. The measured and calculated results qualitatively agree for both polarizations. The calculation was conducted in a nonlocal manner using permittivity ε_T (Eq. 4.2) and ε_L (Eq. 4.3), taking account of gradient of the voltage-induced electron density in the thin layer. Furthermore, we make the following comments on the results.

(i) The thickness of the modulated thin layer was set to 0.4 nm in the calculation, because the thickness yielded the best reproduction of the experimental data.
(ii) Quantitatively, the ER signals were larger in the experiment. This result was attributed to a contribution of bound electrons, which was difficult to evaluate in the permittivity.
(iii) For comparison, a calculated result for p polarization from the ordinary local-response theory using the Drude permittivity is shown with dotted line in Fig. 4.3b; the local-theory result does not reproduce the measured spectrum and exhibits distinct deviation from the nonlocal-theory result.

4.2 Nonlocal Responses in Near-Field Scattering

As plasmonics grew as a research field, the fundamental interest in the quantum mechanical effect on plasmonic resonances was revived. In this section, we mainly address the progress after 2000.

Theoretical implications for nonlocal responses by plasmonic resonators Nonlocal responses of metallic nanostructures attracted interest in accordance with the growth of plasmonics and were theoretically investigated at first for gap plasmons (García de Abajo, 2008; McMahon et al., 2010).

Figure 4.4 shows scattering cross section by local (dashed lines) and nonlocal (solid lines) responses by a Au-nanosphere dimer of gap distant d (García de Abajo, 2008). Each diameter of the nanospheres was set to 20 nm. In case of $d \to \infty$, the two responses result in equivalent cross sections (black curves). As the gap distance d becomes smaller, the discrepancies gradually become obvious. In case of $d < 1$ nm, the cross-section peaks clearly separate. This kind of behaviors is sometimes referred to as red shift by the local response theory, simply because the peaks appear at longer wavelengths.

In the theoretical calculation for Fig. 4.4, quantum mechanical treatment is not the first-principles calculation but a nonlocal-

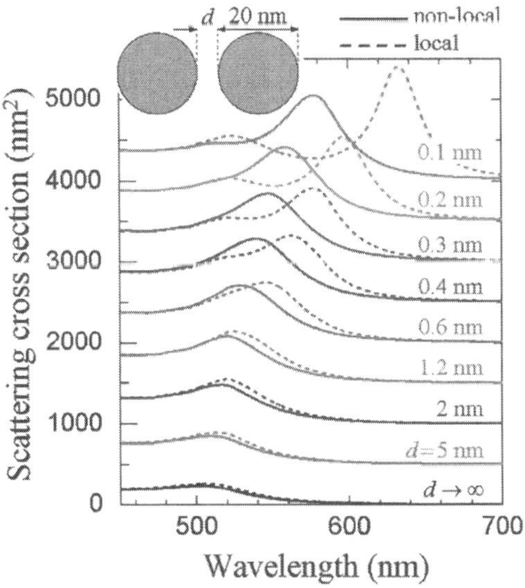

Figure 4.4 Scattering cross section of a Au-sphere dimer. The local (dashed curves) and nonlocal (solid curves) responses are compared varying the dimer gap d from 0.1 to ∞. The local response tends to appear at longer wavelengths than the nonlocal response. The case of $d = \infty$ is equivalent to isolated single sphere. Reprinted with permission from García de Abajo (2008). Copyright 2008 American Chemical Society.

response model (García de Abajo, 2008), which is called specular reflection method. Relative permittivity of homogeneous metal was given by

$$\varepsilon(\omega, k) = \varepsilon_{\text{loc}}(\omega) - \varepsilon_D(\omega) + \varepsilon_M(\omega, k) \tag{4.5}$$

where ε_{loc} is determined by measured values, ε_D is Drude permittivity in Eq. 1.24, and ε_M is dielectric function for longitudinal component, derived based on random-phase approximation and electron-motion damping by collision (Marmin, 1980). Note that the variable in Eq. 4.5 is not wavevector $\mathbf{k}$ but the scalar $k = |\mathbf{k}|$, meaning that the $\mathbf{k}$-dependent response is isotropic. Also, Eq. 4.5 was given in a phenomenological form and the term ε_M was introduced as correction to the difference, $\varepsilon_{\text{loc}} - \varepsilon_D$.

Because sphere functions are generally expanded by the spherical Bessel function j_l, the lth-order polarizability α_l is written, using multipole expansion, as

$$\alpha_l = r^{2l+1} \frac{l(\varepsilon_l - \varepsilon_{\text{host}})}{l\varepsilon_l + (l+1)\varepsilon_{\text{host}}} \tag{4.6}$$

where r is radius of the nanospheres, l is the orbital momentum number, and $\varepsilon_{\text{host}}$ denotes relative permittivity of host material for the Au spheres. The lth-order dielectric function ε_l is analytically given by

$$\varepsilon_l = \frac{\pi}{2r(2l+1)} \left(\int_0^\infty \frac{j_l^2(kr)}{\varepsilon(\omega, k)} dk \right)^{-1}. \tag{4.7}$$

The response from each sphere was described by the multipole polarizabilities. The response from the dimer was self-consistently described as multiple scattering and the scattering cross section was plotted in Fig. 4.4. On the other hand, the local response was evaluated by replacing $\varepsilon(\omega, k)$ in Eq. 4.7 with $\varepsilon_{\text{loc}}(\omega)$. There is a clear tendency that the local response predicts "red" shift; in other words, the cross section peaks evaluated based on the local response appear at longer wavelengths than the peaks based on the nonlocal response. Especially, the discrepancy becomes obvious for the gap $d < 1$ nm; the large discrepancy would be experimentally detected if the Au dimer is precisely prepared.

Experimental examinations for nonlocal responses by plasmonic resonators The revived interest in the limit of the classical regime on plasmonic resonances was stimulated and two independent experiments examining the limit were reported in 2012 (Ciracì et al., 2012; Savage et al., 2012).

Figure 4.5 is one of the experimental results, showing gap-dependent light scattering by a pair of Au tips. The resonant gap modes are understood as gap plasmons between the Au-tip heads, as shown in Fig. 4.5a. The upper panel depicts the two Au tips for atomic force microscope (AFM) were precisely placed at distance d and were illuminated by light through an objective lens. The middle panel shows a dark-field microscope image of the two tips with scale bar (white) of 1 μm. The bottom panel shows a magnified scanning electron microscope (SEM) image of a Au tip; the scale bar (white) indicates 200 nm.

Figures 4.5b and 4.5c show two-dimensional (2D) plots of scattered light intensities in experiment and quantum mechanical

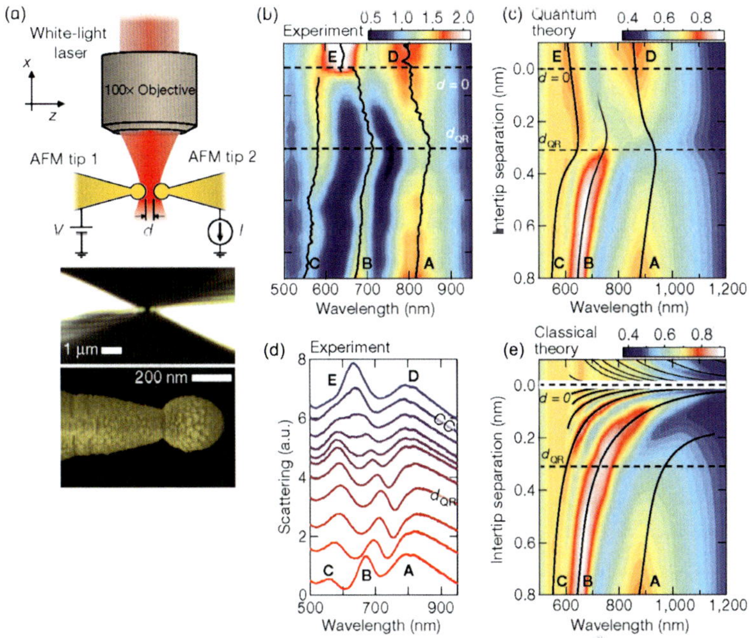

Figure 4.5 Scattering spectra measurement to examine nonlocal responses by gap plasmons. (a) Schematic of the experimental configuration (upper). Two AFM tips form controllable gap of distance d. Photograph of the two tips is shown as the middle panel. SEM image of the tip is shown as the bottom panel. (b) 2D plot of measured scattering spectra represented with pseudo-color. Black curves trace the peaks of the spectra. (c) 2D plot of calculated scattering spectra using the quantum model, shown in a similar way to (b). (d) Measured scattering spectra with various bias voltages. (e) 2D plot of calculated scattering spectra using the classical theory, shown in a similar way to (b). Reprinted by permission from Macmillan Publishers Ltd: (Savage et al., 2012), copyright 2012.

calculation, respectively. The horizontal axes represent wavelength of light, shown in different ranges. The vertical axes represent intertip separation, equal to d. The d was varied from 0.0 to 0.8 nm. Peaks termed A to E are indicated by solid lines. It is evident that qualitative behaviors of the peaks are in good agreement with each other. The quantum mechanical calculation was implemented using a nonlocal-response model, called quantum corrected model (Esteban et al., 2012), which was derived from time-dependent density-functional theory, describing coherent tunneling of electrons. The model is compatible with the classical electromagnetic (EM) description for plasmonic resonance involving a large number of Au atoms on the tips. Thus, the quantum effect is ascribed to tunneling of electrons between the Au tips.

Figure 4.5d shows measured scattering curves, which are plotted in a 2D manner in Fig. 4.5b. The several peaks A to E exhibit nontrivial dependence on d.

Figure 4.5e shows a numerical result based on the classical EM theory. At $d = 0.8$ nm, the peaks A, B, and C appeared, similarly to the experimental result in Fig. 4.5b. However, as d becomes smaller, the behaviors by the classical theory show obvious discrepancies from the experimental results; the experimental peak A turns to the shorter wavelength at a particular d that is indicated by dashed line and termed d_{QR}, and goes to D, whereas the classical calculation could not reproduce it. Similar discrepancies are confirmed as for the peaks B and C. The d_{QR} denotes the boundary of quantum regime (QR). In this experiment, d_{QR} appeared at 0.3 nm, suggesting that the classical calculations for very small scales less than 0.5 nm generally need justification when they are used in researches on plasmonic resonances.

The two experimental results in this heading indicate the limitation of the classical regime for plasmonic resonances. We can summarize the implications as follows.

(i) The classical calculations based on Maxwell equations have the limit on the validity. Plasmonic resonances in metallic nanostructures require special caution when one examine and discuss the effect that originates dominantly from very tiny spatial dimension characterized by a scale less than 0.5 nm. The scale is nearly atomic scale and, in principle, belongs to the quantum mechanical regime.

(ii) In other words, as long as the scales more than 1 nm are mainly studied, the classical calculations provide reliable results that are almost equivalent to the quantum mechanical (i.e., first-principles) results.

Note that, in experiment, the rather broad resonant line width of plasmonic resonances and inhomogeneous broadening of resonant line width coming from fabrication precision are carefully taken into account when spectral shapes are discussed.

4.3 Optical Nonlocality in Plasmonic Resonators

4.3.1 *Linear Response Regime*

As described in Section 4.1, bulk metals themselves exhibit nonlocal responses at the electronic interband transition. In addition, plasmonic resonances induced in the metallic nanostructures also exhibit nonlocal responses as described in Section 4.2. The nonlocal responses explicitly appear in the **k**-dependence of the resonant energies and spectra, due to the response function $\chi(\mathbf{k}, \omega)$.[c] Actual examples have been presented in Chapter 3 (Figs. 3.18, 3.28, and 3.36).

Figure 4.6 schematically shows typical dispersion diagrams in plasmonic resonators. The horizontal axes are taken as good wavenumber vector **k** that characterizes the plasmonic resonators. Most of plasmonic resonators have been produced on substrates or embedded in a layer; therefore, the whole structures are regarded as slabs and then wavenumber parallel to the slab becomes a good wavenumber. The vertical axes represent angular frequency ω, proportional to photon energy. Dispersion diagrams indicate resonance positions in the (k, ω) plane where $k = |\mathbf{k}|$. Plasmonic resonances generally depend on **k** and ω; that is, the response function χ has spatial dispersion such as $\chi(\mathbf{k}, \omega)$. The nonlocality is often referred to as optical nonlocality, which is not limited to the plasmonic resonators but is found in a wide range of artificial nanostructures.

Figure 4.6a shows simply reduced modes (bold curves) into the first Brillouin zone by periodicity a. Diffraction gratings have these modes. Metal-coated gratings also have these modes; then, the modes are SPP-reduced modes because SPP at air–metal flat interface is located close to the light cone (LC, oblique dashed lines) at low frequency ranges. As shown in Section 3.6.5, plasmonic

[c]The relation of the nonlocal response to the **k**-dependence was noted at the beginning of Chapter 2.

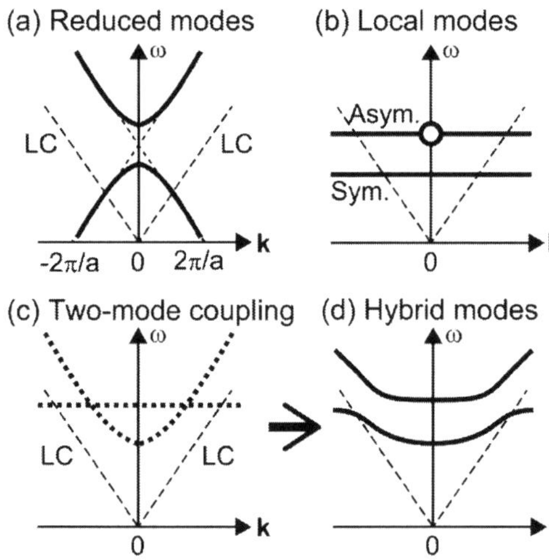

Figure 4.6 Schematic dispersion diagram of plasmonic resonators, which means spatial dispersion, i.e., optical nonlocality. (a) Simply reduced modes in periodic structures of periodicity a. Oblique dashed lines represent light cone (LC). Dotted lines crossing the ω axis at $\omega > 0$ show reduced LC by the periodicity. (b) Local modes independent of wavenumber **k**. (c) Coupling of two different modes is shown with bold dotted lines. (d) After the coupling in (c), hybrid modes shown with solid lines become observable resonances.

waveguide modes are also reduced in a similar manner; the lowest resonance in fishnet metamaterials has been ascribed to the reduced plasmonic waveguide mode and was shown to related to negative refraction effect. Note that the plasmonic waveguide modes in flat metal-insulator-metal structures are usually located below LC (Fig. 3.2).

Figure 4.6b shows local modes (bold lines), which are independent of wavenumber vector **k**. Obviously, the response function χ depends only on ω such as $\chi(\omega)$. Plasmonic resonances in periodic structures are mostly dependent on **k**. Thus, optical nonlocality is a usual property in plasmonic resonators, in contrast to resonances in solids, liquids, atoms and molecules where nonlocality is less frequently observed. One of the exceptions for the optical nonlocality in plasmonic resonators was found in stacked complementary (SC) structures in Section 3.8. The lowest resonance in II-shape SC structure (Fig. 3.41), and the first and second resonances in

SOI[d]-based SC structure are almost independent of wavenumber **k**, irrespective of the periodic structures.

Figure 4.6c displays a situation that two different modes (bold dotted lines) go across with each other. The intersections generally result in qualitative changes of the dispersion diagrams, forming coupled (i.e., hybrid) modes.

Figure 4.6d schematically shows hybrid modes (bold curves) originating from the two modes in Fig. 4.6c. As a consequence of the hybridization, anti-crossing dispersions appear. This is qualitatively understood from the classical two-oscillator-coupling model. This kind of hybrid modes is not limited to plasmon–plasmon coupling. When one of the two original modes is plasmonic resonance and the other has different origin like excitons, so-called strong coupling takes place, as described later (see Section 5.5).

Overall, Fig. 4.6 illustrates that various dispersion diagrams can appear in plasmonic resonators. In other words, it is suggested that the optical nonlocality is observed in diverse shapes of dispersion diagrams.

Comments on the results on plasmonic resonances in Chapter 3

(i) Let us first recall the optical responses by Ag nanodisks in Fig. 3.8 (Section 3.2.3). The absorbance spectrum was computed based on the local responses in the classical electromagnetism whereas, in Chapter 3, it was noted that an experimental data measured in electron energy-loss spectrum (EELS) for the similar Ag nanosphere agreed quite well with the classical calculation result though a small deviation was recognized. The deviation is considered to originate from the difference of the local and nonlocal responses.

(ii) Optical nonlocality has been also shown in stratified metal–insulator metamaterials (SMIMs) (Section 3.6.2). The appearance of transmissive modes in the (k, ω) plane was quite unique as seen in Fig. 3.28. The peculiar optical properties were characterized using $\mathrm{Im}(K_z)$ (for example, Eq. 3.47) where the K_z represents a wavenumber component for the propagation direction. Note that, although plasmonic resonators are mostly

[d]Silicon-on-insulator.

characterized by absorption (or extinction) spectra, SMIMs are characterized by the transmission windows. The photonic Bloch bands form continuum; the TM band originates from SPP at the low frequencies and from quasi-metallic dielectric function (or effective permittivity ε_{eff}) at the relatively higher frequencies near $\varepsilon_{\text{eff}} = 0$.

4.3.2 *Nonlinear Response Regime*

Nonlocal responses in plasmonic resonators are also expected to contribute to nonlinear optical responses in highly efficient and unique ways. This is because higher harmonic generations such as second harmonic generation (SHG) and third harmonic generation (THG) have various (quasi-)phase matching conditions, some of which are distinct from the conditions inferred from the local response regime.

SHG, in principle, appears even in the medium with inversion symmetry if product of E-field vector and spatial derivation of E field is nonzero (Bloembergen et al., 1968). Such a good situation will be prepared based on the designs of resonant EM fields in the optical nonlocal regime.

The third-order nonlinear current density $\mathbf{J}^{(3)}$ in the nonlocal regime was shown in Eq. 2.96, which implies that nonlocality is able to enhance the THG in well optimized configurations of the involved nanostructures. Other various nonlinear optical processes could be observed efficiently.

At present, there have not been many experimental demonstrations along the expectations above because the studies on nonlinear optical responses employing plasmonic resonators have relied on the scheme of simple E-field intensity enhancement and have not been conducted in well optimized manners. Thus, it seems that in-depth designs will open this direction.

4.4 Summary

This chapter first described the boundary of the classical and quantum regimes. As described in Sections 4.1 and 4.2, both single crystals of metal (Ag) and nanostructured plasmonic resonators exhibited nonlocal responses that originate from the quantum mechanical nature. As a result, the limit of the classical regime in plasmonic resonators (i.e., metallic nanostructures) exists at less than 0.5 nm dimensions where the plasmonic resonances are mainly induced. In one of the examples (Fig. 4.5), the characteristic structure was gap between the two metallic nanostructures. This result is understood in the following two ways.

(i) The validity of the classical regime is quite wide and consequently it turns out that there are many plasmonic resonators properly described in the classical regime.
(ii) In contrast, it needs a special care to deal with plasmonic resonances in the classical manner, whose characteristic size is less than 0.5 nm dimensions.

We note that the nonlocal responses were analyzed using not direct first-principles calculation but the quantum mechanical models. In the near future, the first-principles calculations will be able to handle plasmonic resonators directly.

In Section 4.3, optical nonlocality in artificial nanostructures including plasmonic resonators was described. The optical nonlocality directly comes from wavenumber **k**-dependent response function $\chi(\mathbf{k}, \omega)$ and visualized as dispersive diagrams in the (k, ω) plane (for example, Fig. 4.6). Thus, the optical nonlocality was confirmed to be a quite usual property observed in plasmonic resonators (and artificial nanostructures).

References

Bloembergen, N., Chang, R. K., Jha, S. S., and Lee, C. H. (1968). Optical second-harmonic generation in reflection from media with inversion symmetry, *Phys. Rev.* **174**, 813–822.

Ciracì, C., Hill, R. T., Mock, J. J., Urzhumov, Y., Fernández-Domínguez, A. I., Maier, S. A., Pendry, J. B., Chilkoti, A., and Smith, D. R. (2012). Probing the ultimate limits of plasmonic enhancement, *Science* **337**, 1072–1074.

Esteban, R., Borisov, A. G., Nordlander, P., and Aizpurua, J. (2012). Bridging quantum and classical plasmonics, *Nat. Commun.* **3**, 825.

Forstmann, F., and Gerhardts, R. R. (1986). *Metal Optics Near the Plasma Frequency* (Springer-Verlag, Berlin).

García de Abajo, F. J. (2008). Nonlocal effects in the plasmons of strongly interacting nanoparticles, dimers, and waveguides, *J. Phys. Chem. C* **112**, 17983–17987.

Kötz, R., Kolb, D. M., and Forstmann, F. (1980). Evidence for longitudinal waves in electroreflectance spectroscopy of silver, *Surf. Sci.* **91**, 489–498.

Marmin, N. D. (1980). Lindhard dielectric function in the relaxation-time approximation, *Phys. Rev, B* **1**, 2362–2363.

McMahon, J. M., Gray, S. K., and Schatz, G. C. (2010). Optical properties of nanowire dimers with a spatially nonlocal dielectric function, *Nano Lett.* **10**, 3473–3481.

Savage, K. J., Hawkeye, M. M., Esteban, R., Borisov, A. G., Aizpurua, J., and Baumberg, J. J. (2012). Revealing the quantum regime in tunnelling plasmonics, *Nature* **491**, 574–577.

Chapter 5

Plasmonic Enhancement

In this chapter, we survey the application of plasmonic resonators for various enhancement effects. It is described that the diverse plasmonic resonators which appeared in Chapter 3 play key roles in the many effects.

In Section 5.1, the principles of various plasmon-enhanced phenomena (Plas*) are described. We address individual Plas* as follows: concept for artificial control of spontaneous emission rate (Purcell effect) in Section 5.2, plasmon-enhanced photoluminescence (PlasPL) in Section 5.3, surface-plasmon amplified stimulated emission resonator (SPASER) in Section 5.4, strong-coupling of plasmons with excitons in Section 5.5, plasmon-enhanced Raman scattering (PlasRaman) in Section 5.6, plasmon-enhanced catalysis (PlasCat) in Section 5.7, plasmon-enhanced nonlinear optics (PlasNLO) in Section 5.8, and other Plas* except for the above are referred to in Section 5.9.

We also address applications employing metasurfaces of plasmonic resonators at mid-IR ranges in Section 5.10; IR emitters are specified and potential application is referred to. These applications are not always associated with plasmonic enhancement; still, they imply a way to exploit plasmonic resonances.

Plasmonic Resonators: Fundamentals, Advances, and Applications
Masanobu Iwanaga

ISBN 978-981-4745-06-2 (Hardcover), 978-981-4745-07-9 (eBook)
www.panstanford.com

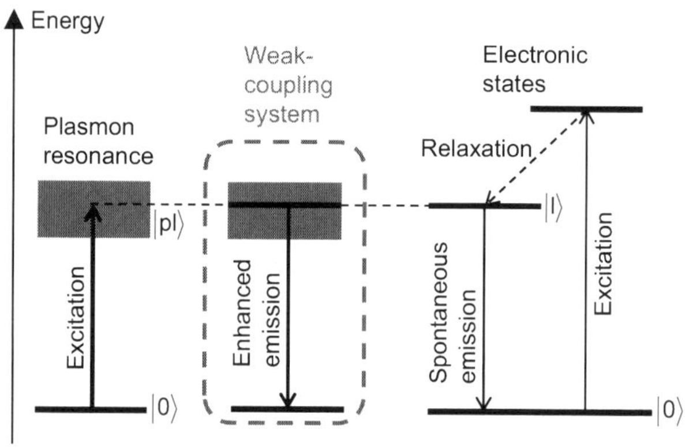

Figure 5.1 Schematic energy diagrams of a Plas* system in weak-coupling regime. PlasPL is taken as an example.

5.1 Principles of Plas*

Weak-coupling regime Figure 5.1 illustrates a Plas* system in weak-coupling regime. As an example, PlasPL is here taken. Two independent resonances originally exist, which are plasmonic resonance and electronic states in molecules, solids, or liquids. The weak coupling is realized by combining the two resonances. However, it is to be stressed that oscillator strength or cross section of the plasmonic resonance is large (bold arrow) whereas transition probability of spontaneous emission in the electronic states is relatively much smaller. In this weak-coupling regime, the plasmonic resonance is hardly affected by the electronic transition; that is, change in the optical spectra of the plasmonic resonance is hardly detected in the coupled configuration. Therefore, the coupling is reasonably termed weak coupling.

In the weak coupling system, the electronic dipole-transition rate Γ is primarily described using Fermi's golden rule (Loudon, 2000) such that

$$\Gamma_{fi} = \frac{2\pi}{\hbar^2}|\langle f|\hat{\mathbf{d}}\cdot\mathbf{E}|i\rangle|^2\delta(E_f - E_i) \tag{5.1}$$

where $\hat{\mathbf{d}}$ denotes dipole operator. The initial state $|i\rangle$ is written as

$$|i\rangle = |l, 0\rangle$$

where l denotes luminescent state and the final state $|f\rangle$ is

$$|f\rangle = |0,\ pl\rangle$$

where pl denotes a plasmonic state of energy E_f. Equation 5.1 is modified (Novotny and Hecht, 2012) such that

$$\Gamma_{fi} \propto \sum_{\mathbf{k},\mathbf{k}'} \hat{\mathbf{d}} \cdot \mathbf{E}_{\mathbf{k}}^{+}\mathbf{E}_{\mathbf{k}'}^{-} \cdot \hat{\mathbf{d}} \exp[i(\omega_{\mathbf{k}'} - \omega_{\mathbf{k}})t]. \tag{5.2}$$

The tensor $\mathbf{E}_{\mathbf{k}}^{+}\mathbf{E}_{\mathbf{k}'}^{-}$ includes information on the E-field intensity dependent on wavevectors $\mathbf{k}$ and $\mathbf{k}'$ where the symbol $\pm$ is associated with annihilation and creation operators. Note that E-field distributions around plasmonic resonators are far from isotropic, as many examples were shown in Chapter 3; therefore, it is improper to assume that the tensor in Eq. 5.2 is isotropic though such simplification has been often conducted.

Decay rate and of emission intensity (i.e., enhanced emission) in weak-coupling systems of fluorescence (FL) molecules and plasmonic resonators are described, based on recent experimental data in Section 5.3.

Strong-coupling regime Figure 5.2 is a schematic illustration of strong-coupling regime. Plasmon and another resonance such as

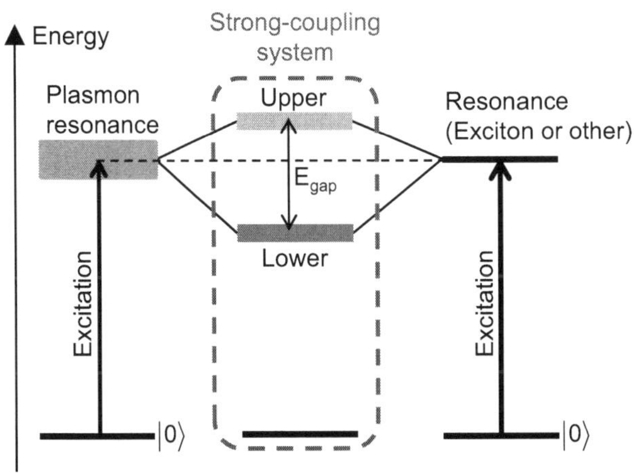

Figure 5.2 Schematic of strong-coupling regime. Plasmon couples with another resonance such as exciton, and the coupling (or hybridization) results in new upper and lower resonances with energy gap E_{gap}.

exciton and waveguide mode couple with each other and form new resonances of upper and lower states. One of the signature in the strong coupling appears as finite energy gap E_{gap}. When dispersion of the upper and lower states is plotted in a relevant plane, anti-crossing dispersion appears.

In comparison with the weak-coupling regime, the strong coupling is made possible only when the two constituent resonances have similar oscillator strength or excitation cross section. In this situation, the two resonances are classically understood as two oscillator with similar resonant energies; then, the two oscillators form a coupled oscillator, which has two levels such as symmetric and anti-symmetric modes. Basically, the strong coupling of plasmon with another resonance is understood in the classical framework.

5.2 Purcell Effect

Original Purcell effect Purcell effect is a base to understand artificial control of spontaneous emission. Originally, Purcell effect was introduced for ideally two level systems; gaseous atoms in microwave cavity were considered and the nuclear magnetic transition rate was estimated for a set of two levels that is resonantly excited at radio frequencies ($\nu = 10$ MHz) (Purcell, 1946). If the atom is in vacuum, the probability of spontaneous emission, so-called Einstein's A factor,[a] is expressed as

$$A_\nu = \frac{8\pi\nu^2}{c_0^3} h\nu \frac{8\pi^3 d_\mathrm{n}^2}{3h^2} \quad (\mathrm{s}^{-1}) \tag{5.3}$$

where d_n denotes nuclear magnetic moment; then, the relaxation time τ $(= 1/A_\nu)$ is approximately 5×10^{21} s. However, if the atom is set in a microwave cavity, the transition rate was significantly enhanced by a factor f such that

$$f = \frac{3Q\lambda^3}{4\pi^2 V} \tag{5.4}$$

where Q signifies quality factor of the cavity, λ wavelength of microwave in vacuum, and V volume of the cavity.

---

[a] As is well known, Albert Einstein considered the coefficients in a classical way (1917). The concrete expression in Eq. 5.3 was derived later, based on the quantum mechanics.

In the original theoretical speculation, it was assumed that small metallic particles of 1 μm diameter were employed to realize the enhancement effect. Then, the volume V is roughly estimated to be $V \sim (10^{-6})^3 = 10^{-18}\,\mathrm{m}^{-3}$, $\lambda = 30$ m, the Q is assumed to be $Q \approx 10$, and finally the factor f results in $f \sim 10^{21}$. A modified relaxation time of $f\tau \sim 1$ s exhibits extreme shortening.

The theoretical prediction was experimentally examined in 1980s, using Rydberg atoms that have transitions at sub-mm ranges. An experiment demonstrated enhancing spontaneous emission rate in Na atoms by employing the cavity resonance at 340 GHz (that is, 0.88 mm in wavelength) (Goy et al., 1983). Another experiment showed inhibition of spontaneous emission of Cs atoms by tuning off the cavity resonance from the quadratic transition (Hulet et al., 1985). Thus, the Purcell effect was experimentally supported in qualitative manners.

At optical frequencies, although luminescence decay rate exhibits significant change in mirror configurations, it is ascribed to configuration interaction of dipole emitters with flat metal (or dielectric) surfaces and should be distinguished from the Purcell effect, as described later (Section 5.3.3).

The Purcell effect was theoretically conceived for an ideal two-level system, which assumed that any other level does not at all contribute to the two-level transition. The ratio of spontaneous emission rates $\gamma_{\mathrm{enh}}/\gamma_0$ (γ_{enh}: the decay rate in enhancing configuration, γ_0: the decay rate in a relevant reference configuration[b]) is usually called Purcell factor, which is an index to quantitatively represent the enhancement effect of the spontaneous decay in artificial platforms.

Purcell effect on artificial nanostructures The Purcell factor was frequently studied in photonic crystals since 1990s because the spontaneous decay rate in photonic crystal cavity attracted much interest. One of the intriguing Purcell effects in photonic crystals was to realize forbidden spontaneous emission in the full photonic

[b]Although vacuum is an ideal configuration, vacuum is not practical configuration for solids and liquids. Therefore, a relevant configuration that does not include any enhancing element is used as an alternative. For example, a transparent media or substrates are often used as reference configurations.

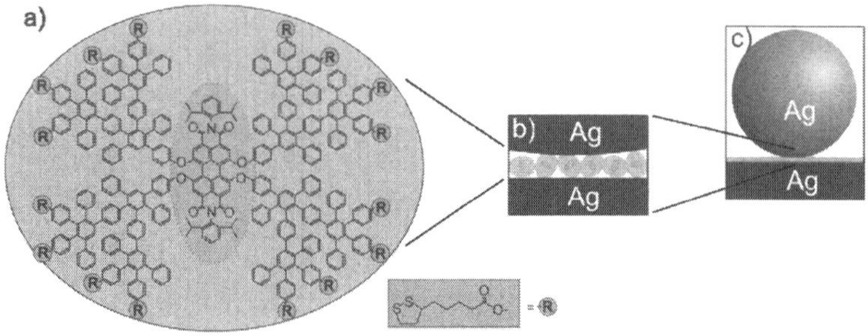

Figure 5.3 A typical FL-enhancing configuration. (a) A dye-loaded dendrimer. (b) Schematic around the Ag gap, in which the dendrimers are located. (c) Whole-view illustration of the Ag-gap plasmonic resonator. Adapted with permission from Schmelzeisen et al. (2010). Copyright 2010 American Chemical Society.

bandgap; then, the decay rate of the spontaneous emission is expected to become very small. Indeed, the corresponding experimental results were reported (Fujita et al., 2005).

Plasmonic resonators also enable to modify decay rates of spontaneous emission. Exploiting the highly enhanced E-field intensity, large FL-decay-rate enhancement up to 1000-fold were observed (Russell et al., 2012; Akselrod et al., 2014). In the experiments, narrow gap plasmon resonators were employed, similar to Fig. 5.3, though the on-top metallic nanostructures were different. Ag nanorods (Russell et al., 2012) and Au nanocubes (Akselrod et al., 2014) were used. Besides, the FL molecules in the gap (or spacer) were different in each experiment.

Relation to FL-intensity enhancement Simple speculation may expect that large FL-intensity enhancement is observed in accordance with the large FL-decay-rate enhancement. This idea is not true because actual spontaneous emissions in FL molecules do not simply come from a two-level transition. In fact, multi-levels including nonradiative (NR) transitions contribute to the FL-related transitions, so that, even when the large decay-rate changes take place, the observed FL intensity seems to be substantially smaller than the expectations (Russell et al., 2012; Akselrod et al., 2014).

Apart from the decay-rate enhancement, FL-intensity or PL-intensity enhancement has been extensively explored for practical potentials. About 40-fold intensity enhancement at the maximum was reported in a semiconductor (Okamoto et al., 2004), whereas more than one-thousand-fold intensity enhancements were recently found for FL molecules (Kinkhabwala et al., 2009; Schmelzeisen et al., 2010; Fu et al., 2010; Zhang et al., 2012; Zhou et al., 2012; Punj et al., 2013; Choi et al., 2015).

Figure 5.3 shows a typical FL-enhancing configuration, based on a Ag-gap plasmonic resonator (Schmelzeisen et al., 2010): (a) FL molecule of a perylene diimide-loaded polyphenylene dendrimer, (b) an extended illustration near Ag gap, and (c) the whole illustration of the gap plasmon resonator, composed of Ag sphere on the plane. From the left to the right, the scale zooms out. In the experiment, the diameters of Ag spheres were in a range from 50 to 100 nm, and the spacer between the Ag gap was 2 $\sim$ 3 nm. Significantly enhanced FL intensity was observed from the top (i.e., above the Ag spheres).

Figure 5.4 shows plasmonic resonators with large and fairly homogeneous FL-intensity enhancement: (a) array of top-Au-dot and bottom-Au-film structure associated with vertical nanogaps (Zhou et al., 2012) and (b) a Au stacked complementary (SC) plasmo-photonic (PlasPh) crystal described in Section 3.8.3.

In Fig. 5.4a, very large FL enhancement was attributed to the nanogaps between the top Au dots and the bottom film; the nanogap is enlarged at the upper right illustration. Self-assembled monolayer (SAM) for human IgG detection is illustrated at the lower left; scanning electron microscope (SEM) images from the top and section views are shown at the lower right. The FL-intensity enhancement factor (EF), defined by ratio of the FL intensity on the vertical nanogap plasmonic resonator to that on reference glass substrate, was reported to be 7400-fold on average. We note that the EF depends on the target molecules themselves as described later (Eq. 5.5); therefore, comparison with other FL molecules does not make sense.

In Fig. 5.4b, optical configuration of incident illumination and FL emission is shown, and very sparsely dispersed molecules are schematically shown (left); it is presented how to disperse the molecule using a pipette on the SC PlasPh crystal that looks blue due to diffraction, having 10-mm square dimension (right). The SEM

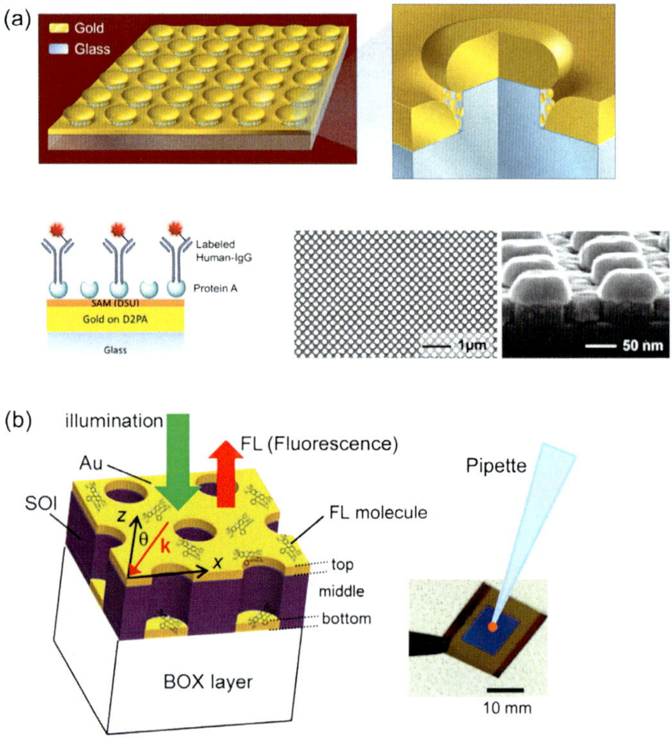

Figure 5.4 Largely FL-intensity-enhancing plasmonic resonators. (a) Array of top-Au-dot and bottom-Au-film associated with vertical nanogaps (upper); SAM-coating for human IgG capture (lower left); SEM images (lower right). (b) Schematic illustration of a SC PlasPh crystal, in which molecules are drawn in a very large scale for clarity. A way of dispersing FL molecules are illustrated at the inset (right). (a) was reprinted with permission from Zhou et al. (2012). Copyright 2012 American Chemical Society. The left of (b) were adapted from Iwanaga et al. (2015) and the right of (b) was adapted with permission from Iwanaga and Choi (2015). Copyright 2015 American Chemical Society.

images of the SC crystal can be seen in Fig. 3.47. A small drop of 2 μL rapidly spread over the substrate and resulted in quite uniform distribution of the FL molecules. Incorporating a SAM on the SC crystal, it was found that thousand-fold FL EF is obtained; the details are described in the next section.

5.3 PlasPL

5.3.1 *PlasPL with Plasmonic Resonators*

PL enhancement by plasmons were already recognized in 1980s (Glass et al., 1980), in which FL was enhanced on Ag islands with incorporating proper spacers. To date, numerous reports were devoted to this subject. Concerning PL of dye molecules or FL, readers can find hundreds of original papers in extensive reviews (Lakowicz et al., 2008; Bauch et al., 2014). As for PL in semiconductors, the studies on PL enhancement are relatively less extensive, probably because the significant effect has not been obtained so often.

Figure 5.5 illustrates a schematic energy diagram involved in the FL enhancement process. The diagram was drawn based on weak-coupling regime of the FL molecules and the resonant EM fields; accordingly, the states are written as $|0, p_i\rangle$, $|l, 0\rangle$, and so on.

The EF in the FL enhancing process (Fig. 5.5) is defined by the FL intensity per unit time and expressed as

$$\mathrm{EF} = \frac{N_{\mathrm{exc}}}{N_0}\frac{\eta}{\eta_0}\frac{\gamma(\mathbf{k})}{\gamma_0(\mathbf{k})} \tag{5.5}$$

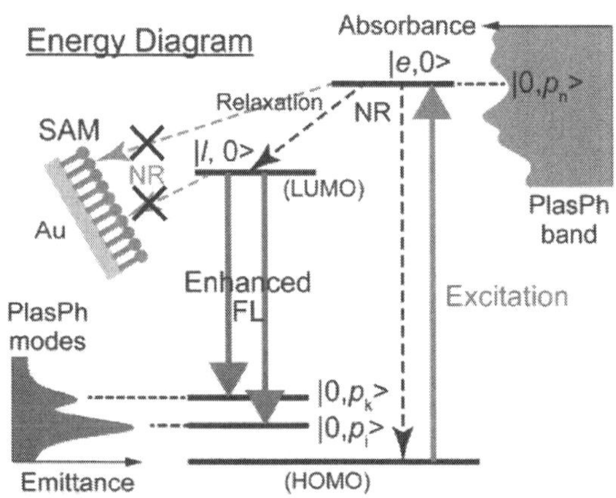

Figure 5.5 Schematic of PlasPL diagram in weak coupling regime of PlasPh resonances and FL molecules. SAM is introduced on the Au surface to enhance FL further. Adapted from Choi et al. (2015) by permission of The Royal Society of Chemistry.

where N_{exc}, η, and $\gamma(\mathbf{k})$ are number of excited states per unit time, FL quantum yield, and radiative decay rate in enhanced circumstance, respectively, and N_0, η_0, and $\gamma_0(\mathbf{k})$ are the corresponding factors in reference circumstance. The symbol $\mathbf{k}$ denotes outgoing wavevector of FL emission.

We here provide the following remarks for Eq. 5.5:

(i) The ratio N_{exc}/N_0 is responsible for excitation efficiency at the incident wavelength, satisfying

$$\frac{N_{\text{enh}}}{N_0} \propto \frac{\sum_m |\mathbf{E}_{\text{enh}}(\mathbf{r}_m)|^2}{\sum_n |\mathbf{E}_0(\mathbf{r}_n)|^2}$$

where $\mathbf{E}_{\text{enh}}$ and $\mathbf{E}_0$ are E field at the positions of molecules ($\mathbf{r}_m$ and $\mathbf{r}_n$) in enhanced and reference circumstances, respectively. The indices m and n run over the involved molecules; therefore, it is important to select the reference that has equivalent amount of involved molecules.

(ii) When NR decay rates in enhanced and reference circumstances are written as γ_{NR} and $\gamma_{0,\text{NR}}$, respectively,

$$\eta = \frac{\gamma}{\gamma + \gamma_{\text{NR}}}$$

and

$$\eta_0 = \frac{\gamma_0}{\gamma_0 + \gamma_{0,\text{NR}}}.$$

(iii) The ratio γ/γ_0 is responsible for radiative decay efficiency, satisfying with

$$\frac{\gamma(\mathbf{k})}{\gamma_0(\mathbf{k})} \propto \frac{\sum_m f(\mathbf{k}, \mathbf{r}_m) \left|\mathbf{E}_{\text{enh}}^{(\text{FL})}(\mathbf{r}_m)\right|^2}{\sum_n f_0(\mathbf{k}, \mathbf{r}_n) \left|\mathbf{E}_0^{(\text{FL})}(\mathbf{r}_n)\right|^2}$$

where factors f and f_0 are wavevector-dependent geometrical factors coming from Eq. 5.2; $\mathbf{E}_{\text{enh}}^{(\text{FL})}$ and $\mathbf{E}_0^{(\text{FL})}$ are E field at the positions of molecules and FL wavelength in enhanced and reference circumstances, respectively.

(iv) Note also that, because EF includes η, EF depends on molecules themselves. Therefore, it is not meaningful to compare different molecules in terms of the EF.

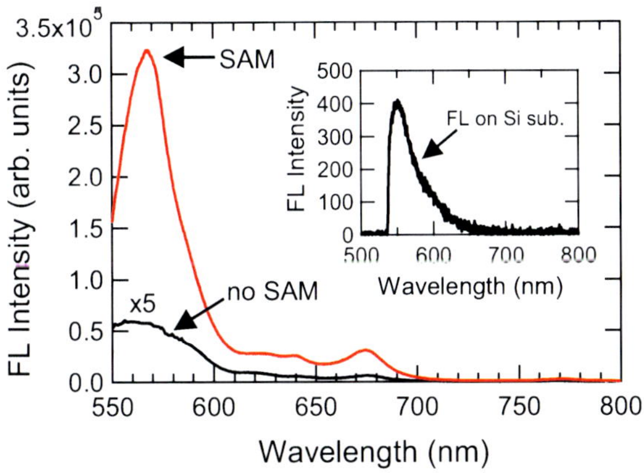

Figure 5.6 PlasPL of R590 molecules with SAM. Red: enhanced FL spectrum on the SC metasurface with a SAM. Black: FL spectra on the SC metasurface without SAM, which was 5 times enlarged for clarity. The inset shows FL spectrum of R590 molecules on a Si wafer; the spectrum was used as reference. Adapted from Choi et al. (2015) by permission of The Royal Society of Chemistry.

Experimental demonstration Figure 5.6 shows a typical enhanced FL spectra that were observed in FL-enhancing platforms. The employed molecules were rhodamine 590 (R590, Exciton Inc., Ohio, USA). Red and black lines represent FL spectra from a SAM-coated SC PlasPh crystals and from a SC PlasPh crystal without any SAM, respectively. Under an equivalent measurement setting, the former is larger than the latter at the peak by more than 30 times. The latter was enlarged by 5 times for clarity. The inset shows FL spectrum of R590 molecules on a reference Si substrate, which was chosen because the absorption capability of the molecules was similar to that of SAM-coated Au films (Choi et al., 2015).

The FL in Fig. 5.6 was induced by a continuous-wave laser at 532 nm at the normal incidence and collected by the objective lens that focus the incidence to a 25-μm-diameter spot. The estimated molecular density on the SC crystals was $1/(35 \times 35)$ molecule/nm^2, implying that each molecule was dispersed isolatedly. The reference FL spectrum in the inset was also measured in the setup using laser-light cut filters and the sharp edge at 535 nm comes from the profile of the filter.

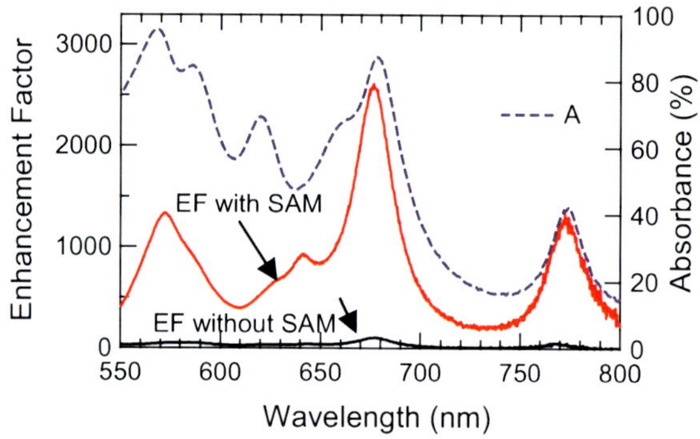

Figure 5.7 EF of R590-molecule FL on the SC PlasPh metasurface with/without SAM. Red denotes EF of FL signals from the SC metasurface with the SAM. Black: EF without any SAM. Purple dashed curve shows A spectrum, plotted for the right axis. Adapted from Choi et al. (2015) by permission of The Royal Society of Chemistry.

Figure 5.7 shows the EF spectra from the data in Fig. 5.6. The EF was evaluated by dividing the enhanced FL spectra with the reference FL spectrum. The largest EF appears at 676 nm, exceeding 2600-fold. Other EF peaks at 570 and 770 nm also reach large values more than 1000-fold. We point out that the EF peaks are mostly consistent with the absorbance (A) peaks. The A spectrum is plotted for the right axis, shown with dashed purple line. According to the theoretical analysis for structured surface (Greffet and Nieto-Vesperinas, 1998), the A is equal to emittance due to the reciprocity; this equivalence is often called Kirchhoff's radiation law. Thus, one of the origins for the significant PL enhancement is ascribed to the high emittance of the SC crystal. E-field enhancement was also estimated to be approximately 50-fold, being one of the factors.

The analysis for the structured surfaces (Greffet and Nieto-Vesperinas, 1998) implicitly assumed the absorption in the constituent materials is small or negligible. The qualitative deviation of the EF from the A spectrum in Fig. 5.7 becomes prominent when the light absorption by the constituent materials of Au and Si is not negligible. We mention that it is, in principle, very difficult to discriminate the absorption by the SC structure and by the constituent materials.

Let us next consider the effect by incorporating the SAM, which was established and known to grow on metallic surfaces in a self-assembled manner (Love et al., 2005). Without SAM, the SC PlasPh crystals exhibit hundred-fold FL-intensity enhancement for the reference Si substrate; these results are shown in Figs. 3.47 and 5.7. By using the SAM of carboxy-end thiolates (DOJINDO Laboratories, Kumamoto, Japan), the maximum EF was 2600-fold for 15-carboxy-1-pentadecanethiol, which was served as a spacer between the molecules and the metallic surface of approximately 1 nm gap. More shorter thiolates were tested, so that 10-carboxy-1-decanethiol was worked similarly while 5-carboxy-1-pentathiol (5-CPT) did not work well for the FL enhancement, probably because the length of 5-CPT is not enough for the Au steps of a few Å. Thus, 20- to 30-fold additional enhancement was experimentally demonstrated owing to the properly chosen SAMs. This SAM-assisted effect was schematically illustrated in Fig. 5.5 as blocking of the NR decay paths.

Summing up the experimental results in Figs. 5.6 and 5.7, we can conclude that the FL-intensity enhancement was made possible from the three main key factors: the high emittance, E-field enhancement, and selection of radiative decay. We also note the followings.

(i) The EF peak and the peak of enhanced FL spectra can always appear at the almost same wavelengths if one tunes the emittance peak of the enhancing platforms to the original FL peak in a reference circumstance.
(ii) The significant EF in Fig. 5.7 does not primarily depend on intense local E fields (i.e., hot spots), so that the EF were observed in a quite uniform way, almost independent of the observing positions on the SC crystals. This property enables to make the detection of the enhanced FL much feasible and will be highly preferred in actual FL-sensing application.

5.3.2 *PL Enhancement in Photonic Crystals*

Recently, non-plasmonic platform was also found to be a good candidate for PL sensing (Choi et al., 2014). It is Si PhCs fabricated in SOI substrates as shown in Fig. 3.53. Figure 5.8a shows measured

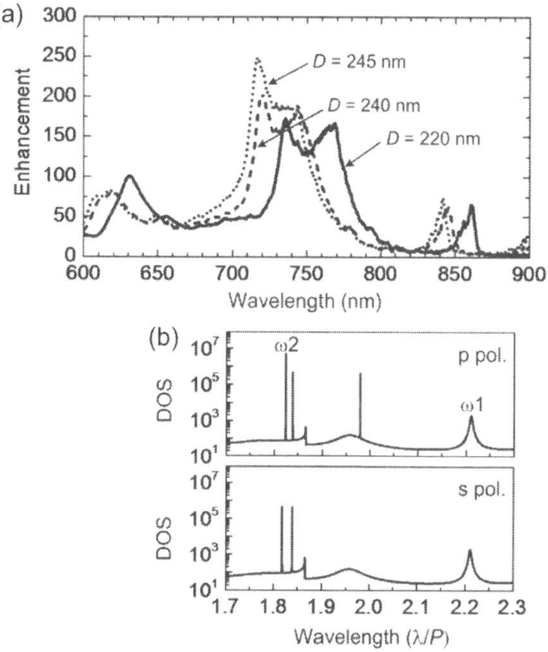

Figure 5.8 PlasPL with Si PhC slabs. (a) Measured EF of R590 molecules on Si PhC slabs of air-hole diameter $D = 220$, 240, and 245 nm. (b) DOS spectra of $D = 220$ nm at p and s polarizations at the Γ point. Adapted from Choi et al. (2014) with permission by AIP. Copyright 2014, AIP Publishing LLC.

EF of R590 molecules dispersed on Si PhCs of air-hole diameter $D = 220$, 240, and 245 nm. It is evident that prominent EFs more than 200-fold were observed. As the D becomes larger the EF peaks shift to shorter wavelengths (i.e., larger photon energies). This shift directly comes from the shift of photonic guided modes in Si PhC slabs.

Figure 5.8b displays calculated density-of-states (DOS) spectra of a Si PhC slab of $D = 220$ nm. Large DOS more than 10^5 is associated with very narrow line width, which means that the modes are high-quality (Q) modes. Note that the DOS spectra are plotted for normalized wavelength λ/P where P denotes periodicity and that the horizontal axes in Fig. 5.8 are displayed to be consistent with each other. An EF peak at about 730 nm in Fig. 5.8a corresponds to the peaks at $\omega 2$ in Fig. 5.8b. Similarly, the EF peaks at 770

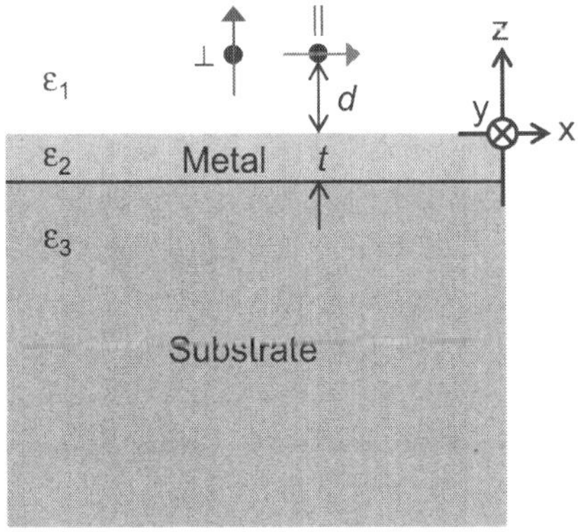

Figure 5.9 Configuration of electric dipoles near flat surface, composed of metallic thin film of thickness t on semi-infinitely thick substrate. The dipoles are assumed to be located at distance d from the metallic surface. Adapted from Choi et al. (2015) by permission of The Royal Society of Chemistry.

and 860 nm correspond to the DOS peaks that have broader line width the peaks that around $\lambda/P = 1.83$. To consider underlying mechanism of the FL enhancement, DOS spectra are suggestive whereas the relation of optical spectra such as R spectra (see Fig. 3.55a) to the mechanism has not been directly accounted for.

5.3.3 *Nonradiative (NR) Decay Rate on Flat Surface*

PL decay time in mirror configurations was examined in 1970s (Chance et al., 1978). We here describe significant increase in NR decay rate in the configuration. Metal-induced FL quenching is often pointed out, which is described as near-field interplay between the dipole and flat surface.

Figure 5.9 illustrates a configuration of electric dipole near flat surface. The dipoles (dots) are located in a semi-infinite domain of permittivity[c] ε_1, which is here assumed to be air, and have two possible orientations, parallel (||) and perpendicular ($\perp$) to the surface. Distance from the dipoles to metallic surface is d and

[c]Relative permittivity is simply referred to as permittivity from now on.

thickness of the metallic film is t. The metallic film of permittivity ε_2 is located on a semi-infinitely thick substrate of permittivity ε_3.

Derivation of damping rates in the flat surface configuration In the configuration of Fig. 5.9, the decay rate of the dipoles are calculated in a classical manner assuming that the dipoles are damping oscillators (Chance et al., 1978). Equation for a damping oscillator has been addressed in Section 2.1 (Eq. 2.12). Here we think of dipole oscillation and write the equation as follows, defining the dipole $\mathbf{p}$ by $\mathbf{p} = e\mathbf{x}$.

$$\ddot{\mathbf{p}} + \gamma_0\dot{\mathbf{p}} + \omega^2\mathbf{p} = \frac{e^2}{m}\mathbf{E} \tag{5.6}$$

Note that γ_0 denotes intrinsic decay rate of the dipole that oscillates at angular frequency ω. The E field $\mathbf{E}$ is formed by emitted field from the dipole in the configuration of Fig. 5.9 and therefore is mainly composed of reflected field from the flat surface.

In general, due to the presence of the field $\mathbf{E}$, the solution of Eq. 5.6 is written as

$$\mathbf{p}(t) = \mathbf{p}_0 \exp[-i(\omega + \Delta\omega)t] \exp(-\gamma t/2). \tag{5.7}$$

Also, the $\mathbf{E}$ has same time-dependence to $\mathbf{p}(t)$ such that

$$\mathbf{E}(t) = \mathbf{E}_0 \exp[-i(\omega + \Delta\omega)t] \exp(-\gamma t/2) \tag{5.8}$$

where $\mathbf{E}_0$ denotes the amplitude at $t = 0$ (i.e., starting time of the oscillation). Substituting Eqs. 5.7 and 5.8 into Eq. 5.6 under ordinary assumptions of $|\Delta\omega/\omega| \ll 1$ and $\gamma/\omega \approx \gamma_0/\omega$, one reaches the following expressions for the frequency shift $\Delta\omega$ and modulated damping constant γ such that

$$\Delta\omega = -\frac{e^2}{p_0 m\omega}\mathrm{Re}(E_0) - \frac{\gamma^2}{4\omega} \tag{5.9}$$

$$\gamma = \gamma_0 + \frac{e^2}{p_0 m\omega}\mathrm{Im}(E_0) \tag{5.10}$$

where p_0 and E_0 denote the complex amplitudes of $\mathbf{p}_0$ and $\mathbf{E}_0$, respectively. Equations 5.9 and 5.10 indicate interesting properties (i)–(iv):

(i) In the flat surface configuration, frequency shift $\Delta\omega$ does appear in the linear response regime and increases in the real part of E_0;

(ii) The damping factor γ always contributes to the shift $\Delta\omega$, which is usually neglected;
(iii) The modulated damping factor γ is dependent on the imaginary part of E_0, which means that purely lossless materials do not induce the modulation;
(iv) In the classical framework, radiative decay constant γ_R is given by

$$\gamma_R = \frac{2}{3}\frac{e^2 k_1^3}{m\omega\varepsilon_1}$$

where k_1 is wavenumber of light in the domain 1. Once we compare the γ_R with Eq. 5.10, it is verified that, in general, the total decay rate γ is not equal to γ_R. Thus, the rate γ includes not only radiative rate but also NR rate.

As for the normalized damping factor $\hat{\gamma} = \gamma/\gamma_0$, it is reasonable to consider the two parallel $\hat{\gamma}_{\parallel}$ and one perpendicular $\hat{\gamma}_{\perp}$ components in accordance with the orientations of the dipoles in Fig. 5.9. Using complex reflectivities r_p and r_s on the flat metallic surface, the components are expressed as

$$\hat{\gamma}_{\parallel} = 1 + \frac{3}{4}q \cdot \mathrm{Im}\left[\int_0^{\infty} \{(1-u^2)r_\mathrm{p} + r_\mathrm{s}\} \exp(2l_1 k_1 d)\frac{u du}{l_1}\right] \quad (5.11)$$

$$\hat{\gamma}_{\perp} = 1 - \frac{3}{2}q \cdot \mathrm{Im}\left[\int_0^{\infty} r_\mathrm{p} \exp(2l_1 k_1 d)\frac{u^3 du}{l_1}\right] \quad (5.12)$$

where the subscripts p and s denote polarizations, and q is intrinsic quantum yield of the dipole, defined by $q = \gamma_R/\gamma$. A variable l_1 means a phase factor of EM wave in the domain of ε_1 and satisfies with $l_1 = -i\sqrt{\varepsilon_1 - u^2}$. Note that the u represents a wavevector component parallel to the surface, normalized by k_0 (k_0: wavenumber of light in vacuum).

Using the $\hat{\gamma}_{\parallel}$ and $\hat{\gamma}_{\perp}$, total decay rate γ_{tot}, which is averaged in an isotropic manner, is written as

$$\gamma_{\mathrm{tot}} = \frac{2}{3}\hat{\gamma}_{\parallel} + \frac{1}{3}\hat{\gamma}_{\perp}. \quad (5.13)$$

Note that, from the definition, the γ_{tot} is also normalized by the intrinsic decay rate γ_0; therefore, if there is no effect on γ_{tot}, the value will be 1.0.

Total decay rate in the flat surface configuration Figure 5.10a shows the total damping rates γ_{tot}, which was directly evaluated from Eq. 5.13 in a numerical way. The emission wavelength was set to 620 nm, which was a FL wavelength of R590 molecules. Solid curves represent the γ_{tot} in the configuration of Au film of $t = 35$ nm thickness on a Si substrate (Fig. 5.9). Dashed curves correspond to the case of $t = 0$ nm (i.e., no metallic film). In both cases, the γ_{tot} significantly increase as the distance d is less than 50 nm. The inset enlarges the range of $d \leq 1$ nm with the log scale for the vertical axis and the two γ_{tot} become very large. The ratio of the two γ_{tot} is almost constant in the range. As mentioned above, when the distance d increases, the effect from the flat surface is expected to be small and indeed the γ_{tot} is close to 1 for $d > 500$ nm. In the numerical evaluations of the decay rates, permittivity of Au and Si was taken from literature (Rakić et al., 1998; Palik, 1991).

Figure 5.10b shows decay times τ, calculated by $\tau = 1/\gamma_{tot}$ and normalized by q. In accordance with the large γ_{tot} at small d less than 50 nm, the τ rapidly decreases. The presentations of curves are in common with Fig. 5.10a.

Figure 5.11 presents an experimental result on the flat surface configuration (Chance et al., 1975). Luminescent Eu ions were embedded in a transparent insulator and set in a layer near the

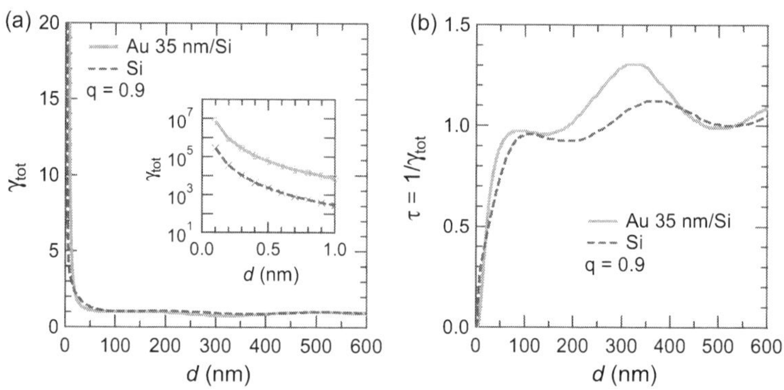

Figure 5.10 (a) Total decay rate γ_{tot} near flat surfaces. Solid curves: γ_{tot} near Au film on Si substrate. Dashed curves: γ_{tot} near Si substrate. (b) Decay time evaluated by $1/\gamma_{tot}$. Adapted from Choi et al. (2015) by permission of The Royal Society of Chemistry.

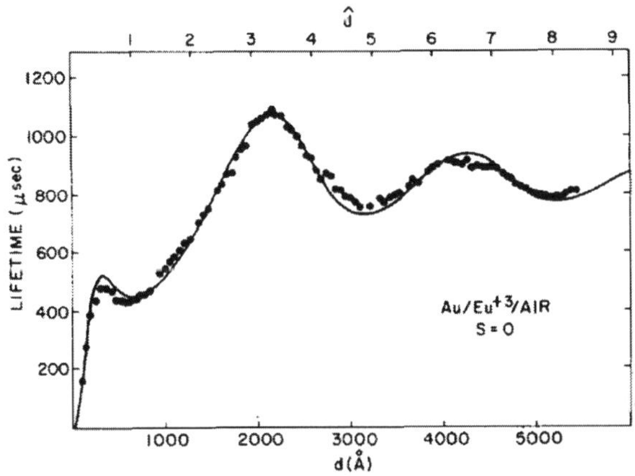

Figure 5.11 Decay time measurement of Eu-ion luminescence. Eu^{3+} ions were located near flat Au surface. Closed circles denote measured data and solid curve represents calculated decay time, similarly to Eq. 5.13. Note that the horizontal axis is in the units of Å. Reprinted with permission from Chance et al. (1975). Copyright 1975, AIP Publishing LLC.

flat Au surface and the distance d was varied from 5 to 550 nm. Measured data are shown with closed circles and numerically evaluated curve is shown with solid curve. The solid curve was calculated in a similar way to Eq. 5.13 and was used to fit the experimental data, resulting in the intrinsic decay time $1/\gamma_0 = 639$ μs and the quantum yield $q = 0.818$ without any flat surface (or mirror). These values are reasonable for the highly luminescent Eu ions. Thus, it was shown that the experiment and numerical calculation well agree with each other.

It is easily verified that the solid curve is quite similar to the calculated curve (yellow line) in Fig. 5.10b. We mention that equation for the total decay rate in the configuration is a little different from Fig. 5.9 because the Eu ions were embedded in a layer grown on the Au surface. Still, the qualitative tendency is similar because it is mainly determined by the presence of the metallic flat surface.

Radiative and NR decay-rate components Figure 5.12a shows a schematic of dipole radiation located near the flat surface and of collection of the radiation using an objective lens. The directions

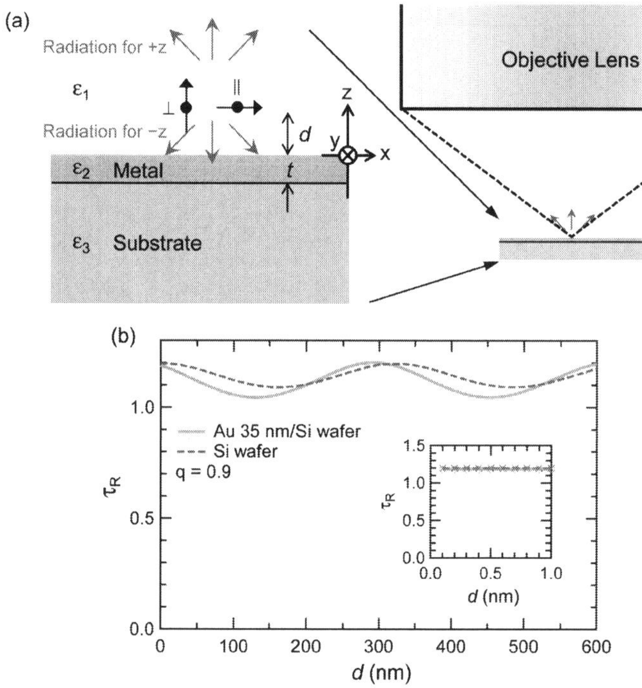

Figure 5.12 Decay rate in an observation configuration. (a) Schematic of configuration. Left: Dipoles near the flat surface. Right: Radiation collection using an objective lens. (b) Plot of radiative decay time τ_R, evaluated by Eqs. 5.16 and 5.18 for comparison with the total decay. Solid curve: τ_R near Au film of 35 nm thickness on a Si substrate. Dashed line: τ_R near a Si substrate. Adapted from Choi et al. (2015) by permission of The Royal Society of Chemistry.

of radiation are classified into two types: one is $+z$-propagating direction and the other is $-z$-propagating direction (Chance et al., 1978).

Taking account of the directions of Poynting vector for the dipole radiation, the total decay rates γ_{tot} can be decomposed into the two parts γ_+ and γ_- where the signs $\pm$ correspond to the $\pm z$ directions, respectively.

$$\gamma_{\text{tot}} = \gamma_+ + \gamma_- \tag{5.14}$$

$$\gamma_\alpha = \frac{2}{3}\gamma_{\parallel}^{(\alpha)} + \frac{1}{3}\gamma_{\perp}^{(\alpha)} + 1 - q \tag{5.15}$$

where $\alpha = +$ or $-$, corresponding to $\pm z$-direction propagation components. Note that we explicitly express intrinsic NR rate $(1-q)$ in Eq. 5.15.[d] Each factor $\gamma_\beta^{(\alpha)}$ ($\beta = \parallel$ or $\perp$) is explicitly written as follows, using n_1 $(= \sqrt{\varepsilon_1})$. The $\gamma_\parallel^{(\alpha)}$ is explicitly expressed as

$$\gamma_\parallel^{(+)} = q - \frac{3}{8}q \cdot \mathrm{Im} \int_0^{n_1} \left\{1 - |r_\mathrm{s}|^2 + \left(1-u^2\right)\left(1-|r_\mathrm{p}|^2\right)\right\} \frac{udu}{l_1}$$
$$+ \frac{3}{4}q \cdot \int_0^{n_1} \left\{r_\mathrm{s} + (1-u^2) r_\mathrm{p}\right\} \exp(2l_1 k_1 d) \frac{udu}{l_1}, \tag{5.16}$$

$$\gamma_\parallel^{(-)} = \frac{3}{8}q \cdot \mathrm{Im} \int_0^{n_1} \left\{1 - |r_\mathrm{s}|^2 + \left(1-u^2\right)\left(1-|r_\mathrm{p}|^2\right)\right\} \frac{udu}{l_1}$$
$$+ \frac{3}{4}q \cdot \int_{n_1}^{\infty} \left\{r_\mathrm{s} + (1-u^2) r_\mathrm{p}\right\} \exp(2l_1 k_1 d) \frac{udu}{l_1}. \tag{5.17}$$

Similarly, the $\gamma_\perp^{(\alpha)}$ components are given by

$$\gamma_\perp^{(+)} = q - \frac{3}{4}q \cdot \mathrm{Im} \int_0^{n_1} \left(1 - |r_\mathrm{p}|^2\right) \frac{u^3 du}{l_1}$$
$$- \frac{3}{2}q \cdot \mathrm{Im} \int_0^{n_1} r_\mathrm{p} \exp(2l_1 k_1 d) \frac{u^3 du}{l_1}, \tag{5.18}$$

$$\gamma_\perp^{(-)} = \frac{3}{4}q \cdot \mathrm{Im} \int_0^{n_1} \left(1 - |r_\mathrm{p}|^2\right) \frac{u^3 du}{l_1}$$
$$- \frac{3}{2}q \cdot \mathrm{Im} \int_{n_1}^{\infty} r_\mathrm{p} \exp(2l_1 k_1 d) \frac{u^3 du}{l_1}. \tag{5.19}$$

Note that the intrinsic quantum yield q is included in $\gamma_\beta^{(+)}$ in accordance with the observation configuration. We note several points, which are made obvious from Eqs. 5.16–5.19

(i) The decay rates $\gamma_\beta^{(+)}$ contain the intrinsic quantum yield q and the integration over a u range of $(0, n_1)$, corresponding to far-field radiation to the $+z$ direction. Therefore, the $\gamma_\beta^{(+)}$ is in connection only with radiative decay rate.

(ii) The NR decay rates coming from the configuration are included in the $\gamma_\beta^{(-)}$. Concretely, the integrals over the $[n_1, \infty)$ range are responsible for the NR decay.

[d]This intrinsic NR rate was implicitly included in Eqs. 5.11 and 5.12. The factor 1 is written as $1 = q + (1-q)$, decomposed into the radiative and NR rates.

(iii) Related to (ii), the origins of the NR decay have not been specified. Therefore, the $\gamma_\beta^{(-)}$ have not lost the generality and any possible contribution is allowed in the $\gamma_\beta^{(-)}$. For example, near-field excitation on the flat surface[e] and energy transfer from the dipole to the metal such as Föster process is included.

(iv) The half of the integrals in Eqs. 5.16–5.19 includes a factor $\exp(2l_1k_1d)$ while the other half does not. This means that the decay-rate components dependent on the distance d are responsible for the near-field excitation and the energy transfer, and that the components independent of d represent far-field modulation for the decay rates. The latter is seen in the sinusoidal variations at $d > 100$ nm, implying an interference effect (Figs. 5.10, 5.11, and 5.12b).

Figure 5.12b shows calculated decay time using Eqs. 5.16 and 5.18. To meet an actual experimental setup such as Fig. 5.12a, we set observable u range to be (0, 0.4), which was determined by the numerical aperture of objective lens in experiment, and implemented the integral over the range. Solid curve represents the radiative decay time τ_R near flat Au film of 35 nm thickness on a Si substrate. Dashed line stands for the τ_R on a Si substrate.

Obviously, the τ_R is not so affected even at small d, as shown in the inset. The sinusoidal variations come from the reason noted just above (iv). Thus, it turns out to be clear that the significant decrease of the decay time at $d < 50$ nm in Figs. 5.10b and 5.11 is ascribed to the rapid growing of NR decay rate. As a result, the total decay rate $\gamma_{tot} \approx \gamma_{NR}$ at the small d. This was a motive to incorporate the SAM in the FL-enhancing experiment (Figs. 5.5–5.7).

We here remark that the radiative decay time in the small d configuration is not detected in experiment because the NR decay rate is dominant. Accordingly, the NR decay time $1/\gamma_{NR}$ is expected to be measured. The extraction of the radiative component in Fig. 5.12b was made possible by the purely theoretical consideration.

[e]For flat metallic surface, plasmonic waveguide modes are excited in the IMI configuration (see Section 3.1.2).

We also remark that the treatment of decay rates has followed a fully classical way, which is expected to be valid for the distance $d \geq 1$ nm from the conclusion in Chapter 4, whereas it would be safe that the decay rates are evaluated in a quantum mechanical way for the very small d less than 1 nm. For example, the decay rates in Fig. 5.10a was possibly overestimated though the qualitative tendency would be similar. To implement the analysis based on the quantum mechanical framework, the method in Section 2.4 and the inclusion of spatial dispersion of metal in Section 4.2 will be useful though the fully quantum mechanical analysis has not been conducted for the configurations such as Fig. 5.9; in the first-principles analysis, the molecule and the atoms near the flat layers are not classical dipoles and have to be fully treated in the quantum mechanics.

Suggestion to the PlasPL on the SC structure In Section 5.3.1, the outmost Au surface was coated by the SAM of at most 1 nm thickness. The classical analysis for decay rate in this subsection implies that the NR decay rate on the reference Si substrate is larger than that on the SAM-coated flat Au surface (see the inset of Fig. 5.10a). Therefore, the FL quantum yield on the reference substrate, η_0, is smaller than that on the SAM-coated Au surface.

Taking account of the experimental result that the ratio γ/γ_0 in Eq. 5.5 has the upper limit of 11 (Choi et al., 2015), the ratio η/η_0 is estimated to be 50 for the EF of 2650, if we assume that the number of involved molecules is almost same in the two configurations. Thus, quite large change of η/η_0 means that NR paths were substantially suppressed by the SAM, as illustrated in Fig. 5.5.

5.4 Surface-Plasmon-Amplified Stimulated Emission Resonators (SPASER)

As an extended effect of PlasPL, stimulated emission is expected. In this section, we survey surface-plasmon-amplified stimulated emission resonators (SPASER), which has potential to realize ultracompact and extremely low-threshold lasers. Although the following examples use different plasmonic resonances such as

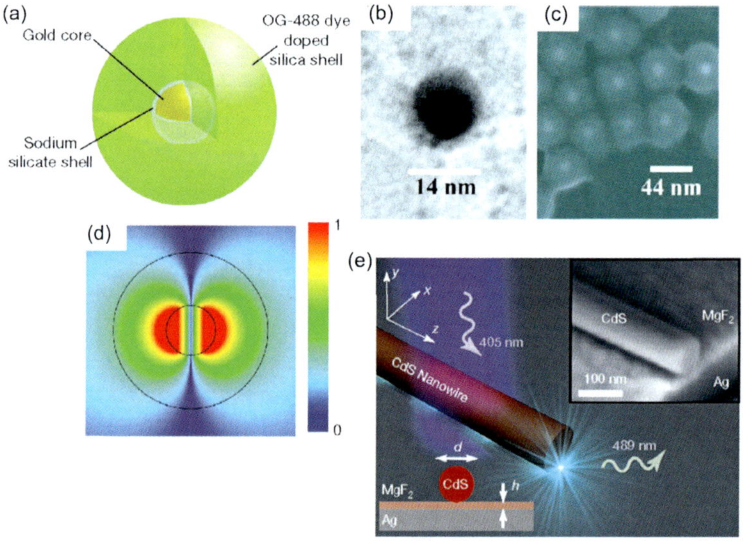

Figure 5.13 SPASER results. (a)–(d) SPASER of core-shell structure. The core was Au nanosphere and the shell was made of silica. The outmost shell was dye-doped. (e) Ag–CdS-nanowire coupled SPASER. Details are given in the text. Reprinted by permission from Macmillan Publishers Ltd: (a–d) (Noginov et al., 2009) and (e) (Oulton et al., 2009), copyright 2009.

Mie-like plasmon, plasmonic standing wave, and plasmonic waveguide mode, we call them SPASER because the plasmon resonances play key role in the enhancement of output light.

Au nanosphere-based SPASER Figures 5.13a–d show Au nanosphere-based SPASER (Noginov et al., 2009). Figure 5.13a shows schematic of the core shell structure. The core was Au nanosphere of 14 nm diameter. The shell was sodium silicate and the outmost shell comprised dye-doped silica. Figure 5.13b presents a TEM image of Au core (black part). The diameter was typically 14 nm. Figure 5.13c displays a SEM image of the core shell structures. The cores (white) are the Au nanosphere and the outer shells are seen around the Au cores. The outmost shells were not always ideally sphere shapes. Figure 5.13d shows numerically calculated E-field strength at the lasing wavelength of 525 nm. The resonant fields appear inside and outside the Au core, being localized in the shell. The quality factor of the resonance was estimated to be 14.8. At the lasing wavelength, Au itself has light absorption due to the interband d–d transition (see Section 1.2.3). It is therefore inferred that excited electrons also contribute, in part, to the excitation of the dye molecules in the outer shell.

Ag-film–CdS-nanowire coupled SPASER Figure 5.13e shows schematic of Ag–CdS-nanowire coupled SPASER (Oulton et al., 2009). CdS nanowires were placed on MgF_2 layer of thickness h; in the experiment, the cases of $h = 5$, 10, and 20 nm were examined. Under the MgF_2 layer, Ag film was prepared in advance. In the case of $h = 5$ nm, the lowest lasing threshold was observed. Standing waves between the CdS nanowire and Ag film played an essential role to the low threshold lasing. Incident wavelength was 405 nm and lasing wavelength was 489 nm. The inset shows a section-view SEM image of the CdS nanowire on MgF_2-coated Ag film.

We note that the lasing itself takes place in the CdS nanowire. The role of Ag film is to induce highly localized standing-waveguide mode (Oulton et al., 2008), which was supported in part by the experimental result that the smaller diameter d of the CdS nanowire resulted in the lower threshold; as a result, the lower lasing threshold was realized. As for the Purcell factor of PL from the CdS nanowires, the factor was at most six-fold and not so large.

SPASER of Ag-surrounding coaxial InGaAsP Figure 5.14a shows a schematic illustration of SPASER of Ag-surrounding coaxial semiconductor (i.e., InGaAsP), shown with red (Khajavikhan et al., 2012). The coaxial waveguide modes in the semiconductor were modified in presence of the metal. In particular, the E-field distributions become strongly localized at the metal/semiconductor interface. The footprint of the SPASER was less than 500×500 nm^2 and has been the smallest to date.

Figure 5.14b displays a transverse-electric-and-magnetic-like coaxial mode and visualizes the strong localization at the interface of the inner metal rod and the semiconductor. The E-field intensity was plotted with color. It is to be noted that, at the low temperature of 4.5 K, the lasing took place at an extremely low threshold, which was almost thresholdless operation.

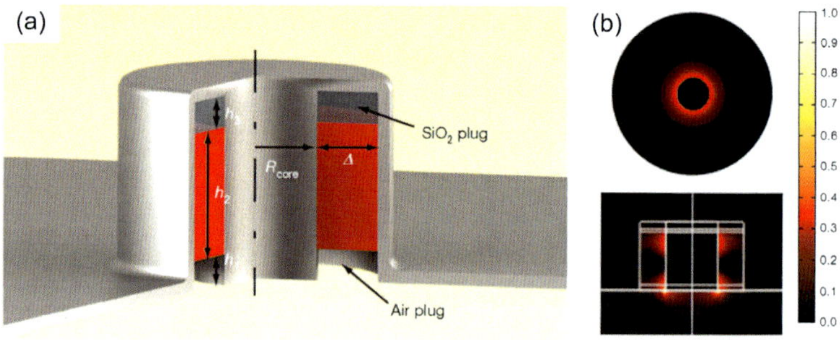

Figure 5.14 Succeeding SPASER result. (a) Schematic of the SPASER structure of Ag-surrounding coaxial semiconductor (red). (b) E-field intensity of a waveguide mode related to the lasing. Details are given in the text. Reprinted by permission from Macmillan Publishers Ltd: (Khajavikhan et al., 2012), copyright 2012.

The three examples of SPASER realized photoinduced lasing in subwavelength volumes. It was consistently shown that more elaborate designs seem to result in lower threshold lasing. Further advances will be expected by conceiving new designs. One of the next challenges will be current-induced lasing.

Comparison with micro lasers of semiconductor Tiny footprint lasers have also been a target in photonic crystals. As an example, it was reported that low threshold lasing at Si photonic crystal cavity of several tens μm dimensions was realized via Raman process and had the threshold for approximately 1 μW input (Takahashi et al., 2013). Contemporary integrated photonic circuits are mainly made of Si waveguides and typically have 100-μm squares and larger dimensions. The micro Si Raman laser has an advantage in the compatibility with the integrated circuits whereas the size will not become smaller as long as the design of the photonic crystal cavity is kept being used.

5.5 Strong Coupling of Plasmons with Excitons and Other Resonances

In a standard theoretical classification, PlasPL in Section 5.3 originates from weak coupling of plasmonic resonances (shortly,

by a both-ends arrow was attributable to Rabi splitting of about 250 meV as a consequence of effective Hamiltonian analysis. The dispersion was a definite signature of strong coupling of the Au-nanorod plasmon with the ITO waveguide modes.

We note that this 1D case is a simpler case of hybridization of plasmons and photonic guided modes than the two-dimensional (2D) SC structures in Section 3.8.3. In the SC structures, plasmons have a variety and large-DOS photonic guided modes contribute to the hybridization, resulting in diverse PlasPh hybrid modes.

5.6 PlasRaman

Raman scattering on metallic surfaces or nanostructures has been extensively investigated. In this section, we address two representative types of PlasRaman studies: surface-enhanced Raman scattering (SERS) and metallic tip-enhanced Raman scattering (TERS). As a related molecular-vibration-enhancing spectroscopy, surface-enhanced infrared absorption (SEIRA) is also addressed in this section.

Figure 5.17 illustrates typical SERS and TERS configurations: (a) A SERS configuration using bare Au sphere-like nanoparticles (NPs, yellow) on a substrate (or support) and probed molecules shown with small dots (pale blue) on the outmost surface of Au NPs; (b) A SERS configuration using Au NPs with transition metal shell (brown) and probed molecules loaded on the NPs; (c) A TERS configuration using a Au tip; (d) An improved SERS configuration using Au NPs isolated with silica or alumina shell on a substrate and probed molecules on the substrate (Li et al., 2010).

We note that there are many other configurations have been reported for SERS and TERS. Let us remind readers that Figure 5.17 is an illustrative example.

5.6.1 *Surface-Enhanced Raman Scattering (SERS)*

Raman scattering on rough metallic surfaces attracted great interest for decades, which is usually called SERS. The enhanced effect was first reported in 1974 (Fleischmann et al., 1974) though, in the

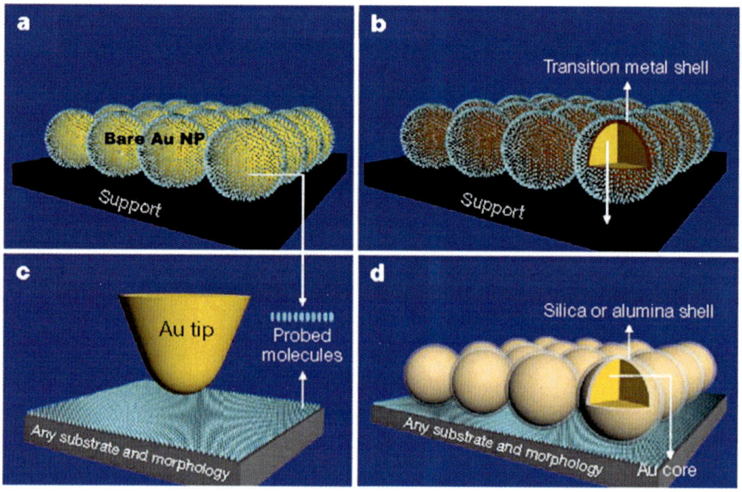

Figure 5.17 Configurations SERS and TERS. (a) A SERS configuration using Au NPs. (b) A SERS configuration using Au NPs with transition metal shell. (c) A TERS configuration. (d) Shell-isolated Au-NP-enhanced Raman scattering. Reprinted by permission by Macmillan Publishers (Li et al., 2010), copyright 2010.

paper, electrochemical effect was mainly addressed and plasmon-enhancement effect was not considered to be a main issue. Three year later, it was recognized that the enhanced Raman signals come from plasmon enhancement (Jeanmaire and van Duyne, 1977; Albrecht and Creighton, 1977).

SERS in 1980s and 1990s Progress in SERS extensively continued through 1980s and 1990s, and the numerous results were reported. Some of the features in the studies are summarized as follows.

(i) Fluorescent molecules such as well-known rhodamine dyes were frequently used as probed molecules. Pump laser light resonantly excited the molecules via the HOMO–LUMO transition. As a result, resonant Raman scattering and FL were simultaneously observed. In the situation, it is, in principle, difficult to distinguish SERS from the electronically resonant Raman scattering. The mixed Raman signals are observable.

(ii) Single-molecule detection was one of the most interested issues. A few papers in 1997 reported single-molecule detection and extremely high enhancement factor up to 10^{14} in the Raman cross Section (Nie and Emory, 1997; Kneipp et al., 1997).

At the time, it was already recognized that purely SERS effect due to E-field enhancement was much smaller than the extreme factor of 10^{14} and that chemical enhancement for the Raman signals substantially contributes to such large enhancement factors (Otto et al., 1992). At the early years in 1990s, metallic nanostructures produced by lithographic method were mainly diffraction gratings; therefore, the E-field enhancement itself was not so large and was estimated to be about 100 fold.

(iii) Metallic NPs were mainly employed for SERS experiment and the hot spots (i.e., intense E-field spots) were assumed to be a main issue to obtain significant SERS signals. However, the hot spots usually appear in the volume smaller than 10 nm^3 and are extremely hard to access. Practically, the hot spots have not been made use. In addition, if one should place the probe molecule at the hot spot, the intense E-field would promptly induce photobleaching and would not allow to obtain the signals from the molecule in a stable manner.

(iv) Metal-induced quenching of molecules was considered to be an obstacle for SERS measurement. As described in Section 5.3.3, NR decay rate dramatically increases on metallic surfaces. Therefore, direct contact of molecules with the metallic surface was highly preferred to be avoided. This was one of the motives in the SERS studies after 2000.

SERS after 2000 Extensive studies on SERS were kept after 2000, let us here refer to a typical result representing recent advances. Figure 5.18 shows high-resolution TEM images of Au NPs with coated with transparent insulator shell. The structure was fabricated to resolve the issue in (iv) just above. Figure 5.18a is a wide view, Fig. 5.18b focus on Au NPs with transparent insulator (SiO_2 or Al_2O_3) shell of different thickness, and Fig. 5.18c magnifies the shells of a few nm, confirming the homogeneous thickness. The scale bars indicate

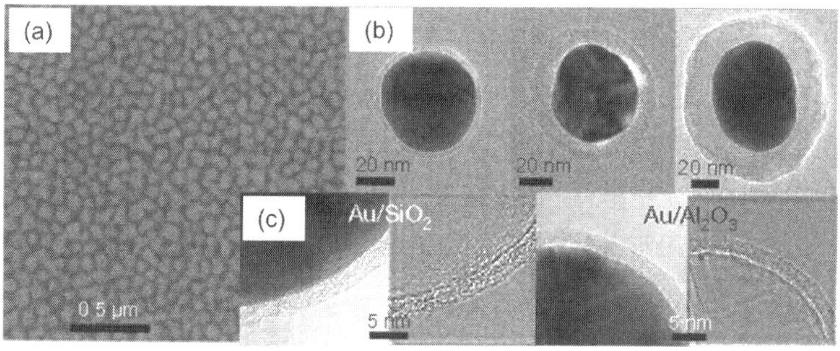

Figure 5.18 High-resolution TEM images of Al_2O_3-shelled Au NPs. (a) A wide view. The scale bar indicates 0.5 μm. (b) Focused views on each NP. The scale bars indicate 20 nm. (c) Focused views on the SiO_2 (left) and Al_2O_3 (right) shell on Au NPs. The scale bars indicate 5 nm. Reprinted by permission by Macmillan Publishers (Li et al., 2010), copyright 2010.

0.5 μm, 20 nm, and 5 nm in Figs. 5.18a–c, respectively. The nm shell was obtained by atomic layer deposition (ALD) technique (Li et al., 2010).

Starting from a motive similar to Fig. 5.18, a different configuration was prepared, which comprised Au NP, thin spacer of insulator, and Au flat film (Mubeen et al., 2012), and significant enhancement of Raman scattering signals was observed up to thousand-fold. Thus, it has become a standard to introduce thin spacer in PlasRaman experiment, which substantially advanced the SERS studies from 1980s and 1990s.

It is interesting to point out that thin films of nm thickness begun to be employed around 2010. An example for FL enhancement appeared in Fig. 5.3. Furthermore, it was recently shown that thin spacer of only 1 nm thickness, which was SAM, has an advantage in the prominent FL enhancement as described in Section 5.3.1. These advances in the Raman-signal and FL enhancements were made possible, thanks to the progress in thin film growth techniques such as ALD (Puurunen, 2005) and SAM (Love et al., 2005).

Figure 5.19a displays Raman scattering spectra in different samples in a practical configuration. Figure 5.19b shows a photograph of the experimental configuration using an orange. The laser light (red) shed on the orange skin from the top (Li et al., 2010).

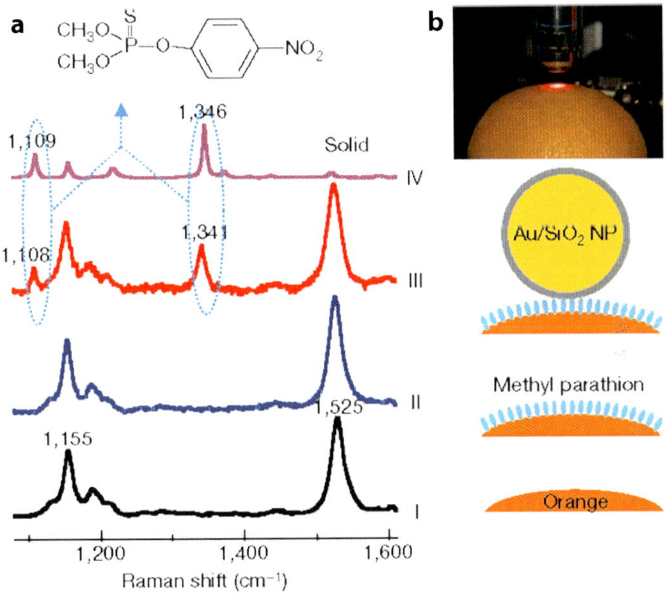

Figure 5.19 Comparison of Raman-scattering spectra. (a) Spectrum I (bottom) shows Raman spectrum from the bare orange skin, spectrum II from the orange skin coated by methyl parathion, spectrum III from the orange skin on which shell-coated Au NPs and methyl parathion were loaded, and spectrum IV from solid methyl parathion. (b) Photograph of the measurement configuration. Reprinted by permission by Macmillan Publishers (Li et al., 2010), copyright 2010.

In Fig. 5.19a, spectrum I (bottom) was measured on orange skin; spectrum II is Raman spectrum of methyl parathion on orange skin without the Au NPs; spectrum III shows Raman spectrum including SERS signals, indicated by dotted-line circles, and was observed using the Au NPs of the shell; spectrum IV is a reference spectrum of solid methyl parathion. The molecular formula of methyl parathion is presented at the top. The key result is found in spectrum III: Only when shell-coated Au NPs were loaded on the orange skin, the Raman signals from the methyl parathion were detected at 1108 and 1341 cm^{-1} (the dotted-line circles). The Raman lines at 1155 and 1525 cm^{-1} come from carotenoid molecules in citrus fruits. The laser power was 0.5 mW on the samples and the spectra were accumulated for 30 s. The setting was typical in conventional Raman measurement. From the signal-to-noise ratio of the measured Raman signals, it does not seem that the enhanced Raman signals are associated with enormous enhancement of the Raman scattering cross section up to 10^{14}, mentioned at the summary (ii) in the previous heading, "SERS in 1980s and 1990s." Instead, the methods enabling to access reproducible Raman signals have been established, which would lead SERS to practical applications.

5.6.2 *Tip-Enhanced Raman Scattering (TERS)*

TERS typically uses metallic tip of nanometer head ($\sim$ 10–20 nm diameter), instead of metallic NPs in SERS. TERS was conceived from a similar idea to SERS and was in connection to the development of scanning probe microscopy since 1980s. In fact, it was reported in 1989 that a small metallic particle that had diameter less than 100 nm was used as a probe in scanning near-field optical microscope (SNOM) (Fischer and Pohl, 1989). The inventions of TERS, which are similar to the metallic-particle-based SNOM in the concept and have differences in the concrete configurations, were independently reported in 2000 by three groups (Stöckle et al., 2000; Anderson, 2000; Hayazawa et al., 2000).

Figure 5.20 illustrates a typical TERS setup. The probe tip has a sharp metallic head and is usually called an apertureless tip. Illumination light comes from the bottom and strongly scattered

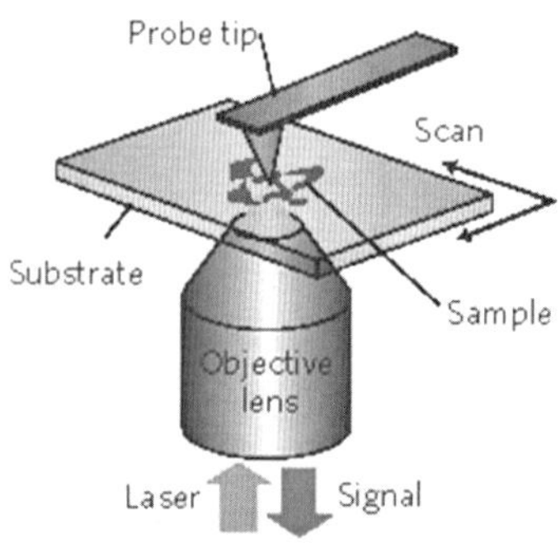

Figure 5.20 Typical TERS configuration. Reprinted by permission by Macmillan Publishers (Kawata et al., 2009), copyright 2009.

light by the metallic tip is also collected by the focusing objective lens. The metallic tip of sharp head, whose diameter is typically 10 nm, locally enhances Raman scattering signals and enables to obtain high-spatial resolution better than diffraction limit. The tip enhancement basically comes from a longitudinal EM mode at the metallic head and possibly allows to access vibration modes of different symmetry from far-field illumination of transverse mode.

We mention that SNOMs of aperture probes were also developed; the probe tip was usually produced based on tapered optical fibers, the head was coated by metal with leaving a tiny aperture of ~20 nm diameter, and light was illuminated and collected through the aperture probe tip.

From the basic setup based on probed microscopes, one of the features in TERS is to be able to measure molecule by molecule, making TERS unique in comparison with SERS. In addition, nanomechanical manipulations are also possible to handle the probe. On the other hand, TERS microscope is a kind of probed microscopy, usually requires high cost, and has a limitation on the throughput in the data acquisition, similar to AFM. At present, TERS microscopes are commercially available through several companies.

5.6.3 *Surface-Enhanced Infrared Absorption (SEIRA)*

Raman scattering detects vibrational signals from molecules or solids as scattered light. SERS in Section 5.6.1 succeeded the enhancement of Raman signals on plasmonic resonators. TERS in Section 5.6.2 uses metallic tip as enhancing plasmonic structures.

Vibrational signals from molecules are also detectable as light absorption spectra in the IR range. It is naturally inferred that absorption enhancement employing plasmonic resonators is possible. Actually, large absorption enhancement at molecular vibrations has been reported to date.

One the examples was carried out in a configuration using metallic rods fabricated on substrate (Neubrech et al., 2008; Adato et al., 2009). This configuration is similar to Figs. 3.14a and 3.15a except that the length of the rods was μm dimensions in accordance with working IR wavelengths. The test molecules for IR absorption were SAM molecules or silk proteins; in both cases, the molecules were directly attached to the metallic rod. From the experimental results, it was concluded that $10^4 \sim 10^5$-fold signal enhancement factors were obtained and that quite small-number molecules were detected. The absorption spectra exhibited Fano-type resonances, described in Section 2.3. In SEIRA configuration, continuum (or broad resonance) is plasmonic resonance and discrete states of narrow resonant line width are molecular vibration states.

Another example combined SEIRA and ultrafast spectroscopy (Selig et al., 2015). The plasmonic resonator was Au metallic rod of μm length fabricated on CaF_2 substrate, which is quite transparent at a wide IR range. Very thin polymethylmetacrylate (PMMA) layer of 5 nm was prepared directly on the metallic rod. The C=O (carbonyl) vibration at 1730 cm^{-1} (i.e., 5.78 μm in wavelength) in PMMA was detected in pump-probe measurement; the measured data represented transient vibration response from the PMMA. The nonlinear response was concluded to be enhanced by 10^4-fold in the presence of the plasmonic resonator.

We note that the prominent two cases addressed here were realized in the configurations that the test molecules were directly attached to or very closely located to the plasmonic resonators. The configurations are a key to obtain the largely enhanced SEIRA signals. This point is in common with efficient SERS and TERS addressed in this section.

5.7 PlasCat

Photocatalysis is currently one of the most extensive research fields in chemistry, aiming at realization of new technologies for energy generation and/or low CO_2 emission. Thus, in addition to purely scientific advantages, economically sustainable technologies are being needed.

Water splitting reaction, $2H_2O \rightarrow 2H_2 + O_2$, by light attracted the first interest in 1970s (Fujishima and Honda, 1972). To date, many attempts were made to improve the conversion efficiency. However, the best efficiency, which is defined by the ratio of the number of generated H_2 molecules to incident photon number, still

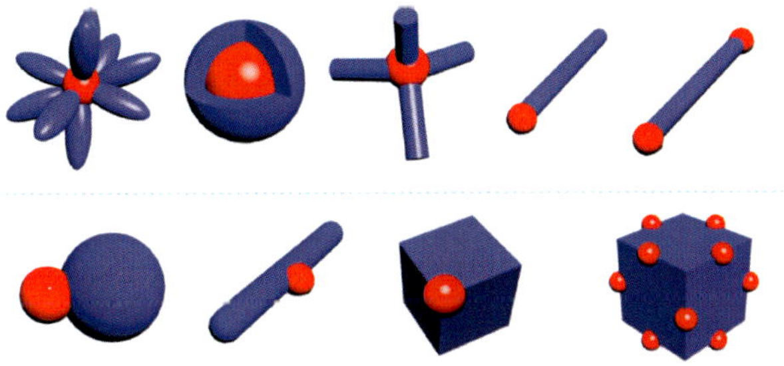

Figure 5.21 Various nanostructures for PlasCat, composed of Au sphere-like NPs (red) and semiconductors (blue). Reprinted with permission from Dutta et al. (2015). Copyright 2015 American Chemical Society.

remains to be around 1% at most. There are numerous reports on photocatalysis using semiconductors such as TiO_2, ZnO, and other complex compounds like perovskite. Recently, the improvement of conversion efficiency and wide working wavelengths are being very vigorously explored (Fujishima et al., 2008; Liu and Chen, 2014).

Let us here focus on plasmon-enhanced catalysis, which we symbolically termed PlasCat. At present, PlasCat is competing with the photocatalysis using only semiconductors.

Strategy to use hot spots Figure 5.21 shows various configurations for PlasCat employing Au NPs (red spheres); various semiconductor nanostructures are illustrated with blue (Dutta et al., 2015). The illustrated structures were experimentally studied so far. We here note that metallic NP-based PlasCat is mainly pushed by chemistry researchers because the metallic NPs are produced through established chemical synthesis and that potential structures are not limited only to the NP-based structures.

Figure 5.22 shows a typical configuration for PlasCat, employing Ag NP cluster (Christopher et al., 2012). The shape of NPs was mostly cube. The NPs were supported on α-Al_2O_3, as shown in Fig. 5.22a.

Figure 5.22b shows extinction spectrum (blue line) obtained from diffuse reflection measurement and emission spectrum from a light source (red line). The extinction and emission spectra have the peaks at 400 and 580 nm, respectively.

Figure 5.22c shows calculated E-field intensity (i.e., $|\mathbf{E}|^2$) of a pair of nanocubes (upper) and of a single nanocube (lower); the gap

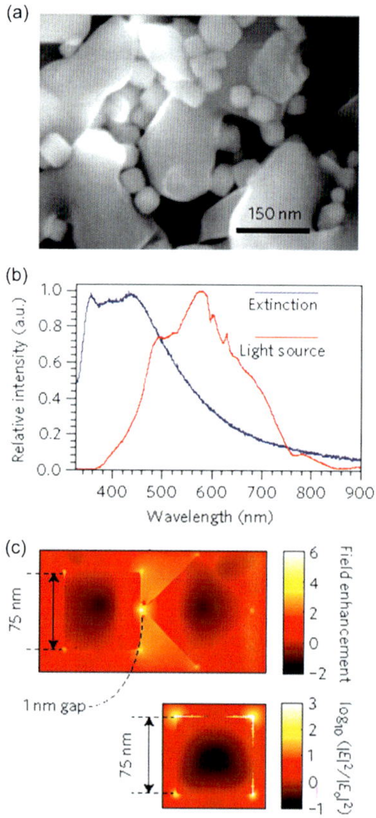

Figure 5.22 Ag NP cluster for PlasCat. (a) SEM image of Ag nanocubes on α-Al_2O_3. (b) Extinction (blue line) and light source (red line) spectra have peaks at 400 and 580 nm, respectively. (c) Numerically calculated E-field intensity at 500 (upper) and 400 nm (lower) by FDTD method. The wavelengths correspond to extinction peaks in each structure. Reprinted by permission by Macmillan Publishers (Christopher et al., 2012), copyright 2012.

of the two Ag nanocubes at the upper panel was set to be 1 nm. Note that the intensity distributions are plotted in a logarithmic scale of $\log_{10}(|\mathbf{E}|^2/|\mathbf{E}_0|^2)$ where $\mathbf{E}_0$ denotes incident E-field vector. The hot spots reach the 10^3- to 10^4-fold intensity for the incident light and were expected to serve as highly efficient reaction points. The strategy to use hot spots is similar to that in the early stage of

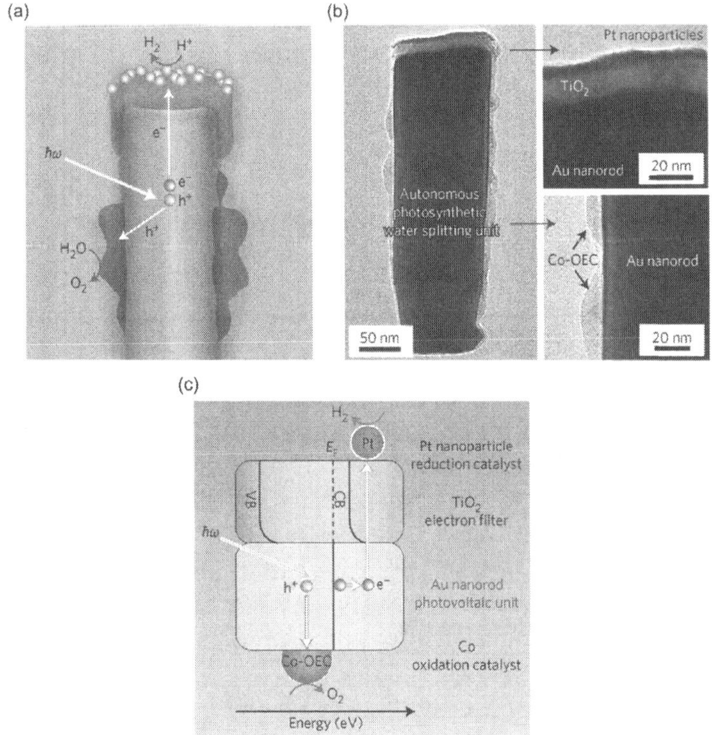

Figure 5.23 Au nanorod for PlasCat. (a) Schematic of PlasCat process in the Au-nanorod system. (b) TEM images of an Au nanorod, corresponding to (a). (c) Schematic of the catalysis mechanism in the Au nanorod. Reprinted by permission by Macmillan Publishers (Mubeen et al., 2013), copyright 2013.

SERS and TERS. The computation was implemented using the FDTD method.

Strategy to use hot electrons Figure 5.23a shows a PlasCat configuration employing Au nanorods on which H_2- and O_2-gas generation parts are loaded. The H_2 evolution was ascribed to Pt nanoparticles deposited on thin TiO_2 layer that was grown by ALD technique on the Au nanorod and designed to be used only as an electron filter. The O_2 evolution was attributed to cobalt-based oxygen evolution catalyst (Co-OEC) grown directly on the Au nanorod. Excited electrons were considered to go through the thin TiO_2 and to reach Pt nanoparticles on the TiO_2 layer.

Figure 5.23b displays TEM images corresponding to the schematic in Fig. 5.23a. The magnified images around Pt nanoparticles on TiO_2 and Co-OEC are shown at the right-hand side. The diameters of the Pt nanoparticles were less than 5 nm and the thickness of the TiO_2 layer was approximately 10 nm, whereas the thickness of the Co-OEC was less than 10 nm.

Figure 5.23c illustrates the mechanism of water splitting in the Au-rod unit. The Au rods were excited by visible light; therefore, it was inferred that hot electrons were induced in the Au rod via the *d*–*d* interband transition and that local plasmons in the Au rod hardly contributed to the chemical process. After the photoexcitation, it was considered that the hot electrons selectively went through the TiO_2 layer, reached the Pt NPs, and resulted in the H_2 evolution. The holes were considered to transfer to the Co-OEC and to induce the O_2 evolution. We point out again that, in this study, the role of Au rod is not a plasmonic resonator but a source of hot electrons. Note also that TiO_2 is employed as an electron filter, not as photocatalysis material.

Hot (or excited) electrons in metal were also discussed in the system of Fig. 5.22; the superlinear catalysis-reaction rate for incident light intensity was attributed to the hot-electron contributions at the hot spots (Christopher et al., 2012). Thus, beyond simple plasmonic E-field enhancement, some electronic effects have been recently investigated.

In spite of very extensive studies on H_2- and O_2-evolution photocatalysis over 40 years, including PlasCat, one of the critical points, that is, the efficiency of conversion rate does not seems to substantially advance. The challenge to pursue the significant efficiency is still an open issue.

5.8 PlasNLO

Surface SHG on flat metallic surfaces Nonlinear optics were launched in 1960s (Bloembergen, 1996; Shen, 2003).[h] At the time,

[h]The original editions were published in 1965 and 1984, respectively.

second harmonic generation (SHG) from flat metallic surface was also analyzed together with other media with inversion symmetry (Bloembergen et al., 1968). Generally, electric quadrupole and magnetic dipole terms contribute to SHG in the perturbation theory.

At the flat surfaces, the two following situations were considered in accordance with the experimental condition.

(i) If one thinks of a low-frequency range that no electronic excitation takes place, electrons in the media are considered to be bound in the orbitals. The quadrupole contribution $\mathbf{P}_{\mathrm{NL,Q}}$ is expressed as

$$\mathbf{P}_{\mathrm{NL,Q}} \approx c_1(\omega)\mathbf{E}(\omega)\{\nabla \cdot \mathbf{E}(\omega)\}.$$

(ii) Conduction electrons in the metal are approximated to be classical plasma at the low-frequency range. The electrons were described using hydrodynamic equation of motion. The nonlinear polarization $\mathbf{P}_{\mathrm{NL}}$ is finally written as

$$\begin{aligned}\mathbf{P}_{\mathrm{NL}}(2\omega) = c_1(\omega)\mathbf{E}(\omega)\{\nabla \cdot \mathbf{E}(\omega)\} + c_2(\omega)\{\mathbf{E}(\omega) \cdot \nabla\}\mathbf{E}(\omega) \\ + c_3(\omega)\nabla\{\mathbf{E}(\omega) \cdot \mathbf{E}(\omega)\}.\end{aligned} \quad (5.20)$$

We note that the coefficient c_1 is proportional to the square of linear susceptibility, $\chi_{\mathrm{L}}(\omega)^2$, and that others c_2 and c_3 are dependent on material parameters in the media. Equation 5.20 includes $\mathbf{P}_{\mathrm{NL,Q}}$ and is a general expression under the low-frequency assumption. We mention that Eq. 5.20 is modified to the following:

$$\begin{aligned}\mathbf{P}_{\mathrm{NL}}(2\omega) = c_1(\omega)\mathbf{E}(\omega)\{\nabla \cdot \mathbf{E}(\omega)\} + \tilde{c}_2(\omega)\{\mathbf{E}(\omega) \cdot \nabla\}\mathbf{E}(\omega) \\ + \tilde{c}_3(\omega)\mathbf{E}(\omega) \times \mathbf{H}(\omega),\end{aligned} \quad (5.21)$$

in which magnetic dipole contribution $\mathbf{E} \times \mathbf{H}$ appears in an explicit form. The coefficients $\tilde{c}_2$ and $\tilde{c}_3$ were redefined according to the modification.

Comparing the measured surface SHG from Ag, Au, Si, and Ge, the metals exhibited larger SHG signals than Si and Ge (Bloembergen et al., 1968). Later, the contributions of the SPPs at the flat metallic surfaces were explicitly observed in the experiment using an attenuated total reflection (ATR) prism (Simon et al., 1974) as shown in Fig. 5.24. The inset draws the ATR and incident configuration, which is similar to Fig. 1.10. The ATR prism enabled to

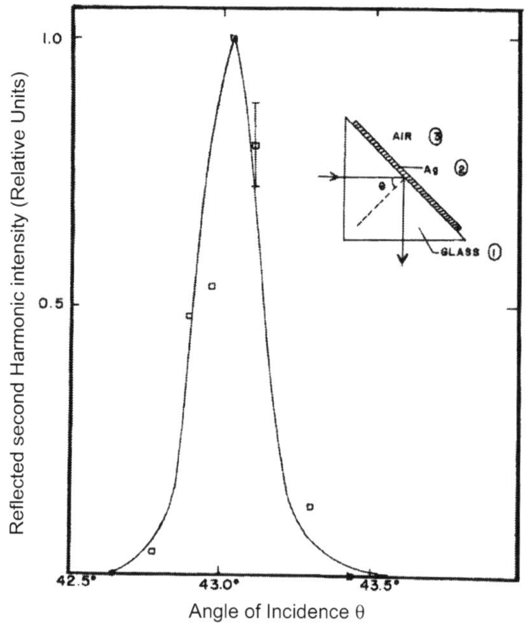

Figure 5.24 Measured SH intensity dependent on incident angles in an ATR configuration. The inset draws the setup for SHG. Reprinted with permission from Simon et al. (1974). Copyright 1974 by the American Physical Society.

excite SPP at the Ag/glass interface at the incident angle $\theta = 43.1^\circ$. A prominent peak of reflected second harmonic (SH) intensity was observed.

Ionized-gas-assisted 17th-order harmonic generations in a plasmonic resonator One of the prominent results in PlasNLO is higher harmonics generation up to the 17th order, which was observed from ionized Argon (Ar) gas at the hot spot of Au bowtie antenna (Kim et al., 2008).

Figure 5.25a schematically illustrates the experimental configuration and the underlying mechanism of the higher harmonics generation. Ultrafast laser pulses of 10 fs from a mode-locked TiS oscillator was injected into a vacuum chamber and focused on a nanostructure, which was Au bowtie antenna array, using a focus lens. Simultaneously, Ar gas jet was shed on the focal spot as

Let us here mention the following two examples showing enhanced light emitters though there are many other reports with similar motives.

A trial was conducted for InGaN-based multi-quantum well structures (Okamoto et al., 2004). To obtain light-emitting enhancement at blue wavelengths, metallic (i.e., Ag, Al, and Au) film and nanostructures were employed. Due to the high-scattering efficiency of the NPs, the net light-emitting efficiency was enhanced by about one order of magnitude.

A different design to use a single-layer plasmonic resonator of Al-disk array was tested for enhancing light emitting (Lozano et al., 2013). The Al-disk array was similar in structure to Fig. 3.8a. The Al-disk array was embedded in a 650-nm thick layer of dye-doped polymer matrix. The light emission was enhanced by 60-fold at the normal direction, suggesting the potential to actual LED application.

The two examples showed that the plasmon enhancement play positive roles. Still, it seems that plasmonic resonators have not been implemented in commercial LEDs to date. This may come from the restrictions that, in the practical stage, the longtime stability and the cost effectiveness are needed to be proved.

Plasmon-enhanced PV devices PV device is one of the most extensively investigated devices. The attempts to increase in the efficiency have been made by many researchers. As one of them, plasmon-enhanced PV devices attracted interest (Atwater and Polman, 2010).

One of the scenarios was to incorporate metallic NPs (or nanostructures) in the PV devices and to induce large light absorption even in thin PV devices. However, there is an inevitable issue that the absorption takes place dominantly at the metallic NPs (or nanostructures) and does not at all ensure excitation energy transfer to the PV structures. Indeed, the net plasmon enhancement for PV devices has not become prominent.

As we have noted already, the control of excited states is generally difficult and requires substantial mechanism to manipulate them. Such manipulation mechanism has not seemed to be realized so far.

Plasmon-enhanced photodetectors Perfect absorbers were described in Section 3.9, implying that light energy is *trapped* at the plasmonic resonators. Therefore, it is expected to be able to convert the trapped energy into current. The PV devices in the preceding heading were studied with a similar motive.

An experimental demonstration for the plasmon-enhanced photodetectors was reported (Knight et al., 2011). The Au nanorods were fabricated on n-type Si that is an electrode. Typical Au-nanorod structure on substrate was illustrated in Fig. 3.14b. A SiO_2 layer was located on the n-type Si and around the Au nanorods. An ITO layer, which was the other electrode, was made on the SiO_2 and Au nanorods; the SiO_2 layer electrically separated the ITO and n-type Si. NIR light illumination yielded photocurrent; in particular, it was stressed that the light energy was below the bandgap energy of the Si. The photocurrent was measured in agreement with extinction cross section of the Au nanorods. This example showed that plasmonic resonators are capable of yielding photocurrent even for sub bandgap light.

Plasmon-enhanced photodetectors reported so far have been limited to a small number. Considering the sub bandgap operation, it seems possible to develop efficient photodetectors not only at the visible but also at IR ranges. Of course, it is necessary to show better performance in comparison with the existing photodetectors.

5.10 IR Emitters

IR emitters attracts substantial interest as application of metasurfaces. In the IR emitters, plasmonic enhancement is not explicitly made use of. Instead, plasmonic resonators make it feasible to design and tune the IR emitters to particular working wavelengths.

5.10.1 *A Practical mid-IR Thermal Emitter*

Figure 5.27 shows a typical mid-IR emitter implementing a plasmonic resonator. The design of plasmonic resonator was same to patched MIM structure in Fig. 3.58. Such patched MIM structures have been frequently used to produce IR emitters (Puscasu and

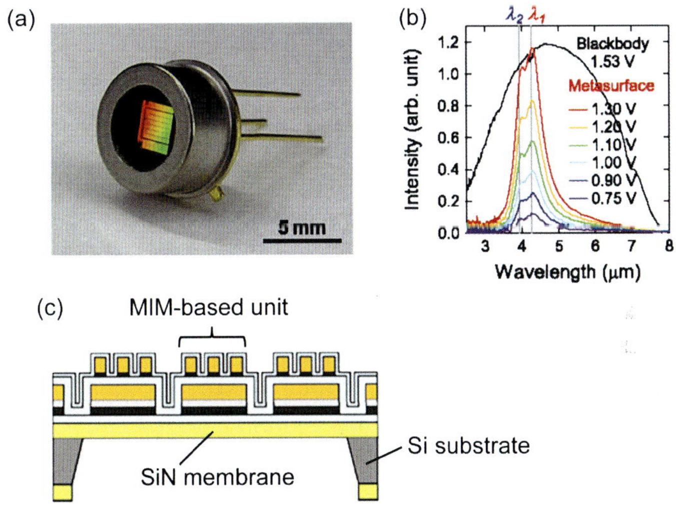

Figure 5.27 Packaged IR emitter based on MIM-based metasurface. (a) Photograph of the whole view. (b) Measured IR emission spectra. Black line denotes blackbody radiation shown for reference. Colored lines show IR emission spectra, which have two peaks at λ_1 and λ_2, from the metasurface emitter for various applied voltages from 0.75 to 1.30 V. (c) Schematic of the section-view IR emitter. Adapted from Miyazaki et al. (2015).

Schaich, 2008; Ye et al., 2008) because perfect absorbers (PAs) are designed with a small effort (see Section 3.9).

The MIM-structure-based design in Fig. 5.27a was realized in a practical manner using high-throughput UV NIL (Section 1.6). The total thickness of plasmonic resonator was approximately 160 nm and the working wavelengths were 4 μm; therefore, the deep subwavelength thickness enables us to call the MIM-based plasmonic resonator metasurface. The IR emitter in Fig. 5.27a was produced to conduct *in-situ* detection of CO_2 gas with 30% lower power consumption than conventional commercial IR emitters (Miyazaki et al., 2015).

Figure 5.27b presents measured IR spectra. The IR emission from the metasurface has two peaks at λ_1 (4.26 μm) and λ_2 (3.95 μm), plotted with colored lines in accordance with applied voltages from 0.75 to 1.30 V. The maximum voltage made the IR emitter to be 305°C. As reference at 305°C, blackbody radiation appears at a mid-

IR range and forms the broad distribution, shown with black line. We note that the λ_1 is an absorbed line by CO_2 molecules and the λ_2 is used as reference for the in-situ detection. The IR emission band was precisely controlled by adding the covering Al_2O_3 using ALD technique.

Figure 5.27c illustrates a section of the metasurface IR emitter. The MIM-based unit works as a plasmonic resonator as shown in Fig. 3.58. The metal M was Au and the insulator I was Al_2O_3. To avoid thermal diffusion, it is important to make the whole structure to be membrane. Thus, thick Si substrate (gray) below the SiN membrane was removed in the fabrication procedure.

Comparison with photonic crystals (PhCs) IR emitters have been also a subject in PhCs and were produced using photonic structures such as gratings since 1980s (Hesketh et al., 1986). PhCs launched in 1986 and 1987 (Ohtaka and Tanabe, 1986; Yablonovitch, 1987; John, 1987) and the development made control of IR emission spectra much sophisticated. Even at high temperatures over 1500 K, it was reported that tungsten (W) PhC works at a near-IR range (Lin et al., 2003).

A recent progress in PhC emitter is a wavelength-selective, high-repetition IR emitter at 9.3 μm, realizing 100 kHz operation without decreasing the normalized emission power (Inoue et al., 2014); the IR emitter was based on the control of intersubband transitions in multi-quantum well of semiconductors and made use of emissive modes of a PhC.

Let us discuss the IR emitters based on plasmonic resonators and PhCs. The wavelength-selective IR emission has been realized in both structures. PhC-based IR emitters are able to emit relatively narrower lines than plasmonic IR emitters. The choice of the two types of IR emitters would depend on application; in some applications, narrow emission lines are necessary and would need PhC-based designs, while, in other applications, rather broad and intense emissions are preferred and then plasmonic resonators would be suited. High-repetition operation up to 100 kHz is at present possible only by PhC-based IR emitters; therefore, in application working at high-repetition rate, PhC-based IR emitters will be a solution. Of course, practical application always requires low-cost production and long-time operation; these points should be also considered in reality.

Recently, reuse of heat attracts great interest to generate electric power irrespective of day and night. Heated reservoirs spontaneously emit IR light, which can be used as energy source for IR photovoltaic devices. The IR source is essentially broadband and therefore broadband photovoltaic devices are preferred. In this application, two options using PhC-based and plasmonic-resonator-based designs are being explored.

5.10.2 *Roles of Plasmonic Resonators at mid-IR Range*

Related to the mid-IR emitters,[i] let us consider possible roles of plasmonics at mid-IR ranges. As described so far, plamonic resonators are able to efficiently absorb EM waves irrespective of the wavelength ranges and to transduce photon energies to plasmonic resonances, that is, local collective oscillation of electrons (Section 3.9). Taking account of the above, IR detectors and sensing applications to biological and chemical objects are easily inferred. Considering the reverse process of light absorption (i.e., reciprocity), IR emitters are also conceived with ease. One of the realization has been shown in Fig. 5.27.

Qualitatively, the following points are more feasible at the mid-IR ranges than the visible range.

(i) Many metals employed in plasmonic resonators become quite good conductors at the ranges, so that the options to select metals become rich. Al, Cu, Ni, and others can be used instead of Au and Ag, as shown in Figs. 1.5 and 1.6.
(ii) The unit dimensions of the artificial structures become larger and often reach μm dimensions. Therefore, photolithography is expected to be used as a key fabrication technique, making the production low cost and mass-productive.

We note that dielectric and semiconductor materials should be chosen with care because optically transparent materials such as SiO_2 and Al_2O_3 are not always transparent at the mid-IR ranges

[i]We here define a range around 2–5 μm range to be mid-IR range.

because the phonon bands often appear at the frequency range. From the optical constants, potential materials are considered to be basically Si and Ge, and potential materials to extend the transparent window at IR ranges are the alloys such as SiGeSn (Soref, 2010). If one sets the base on the IV materials for IR photonics, fabrication techniques and designs can be mostly inherited from Si photonics at the NIR range.

As for light sources, thermal IR emitters and quantum cascade lasers are available at the IR ranges. Therefore, the role of plasmonic resonators at the IR ranges will be mainly enhancement of electronic or vibronic transitions; in fact, SEIRA in Section 5.6 accessed molecular vibrations at IR ranges.

Also, plasmonic resonators serve as to make the IR devices compact, as subwavelength optical devices have been shown at the visible and NIR ranges in Sections 3.6.4 and 3.8.2.

5.11 Summary

In this chapter, we first described the principles of Plas*; then, we described and reviewed various plasmonic enhancement. In comparison with the initial plasmon-enhanced studies in 1980s and 1990s, the configurations of materials (often, molecules) and plasmonic resonators have become more elaborate in 2010s. As a result, the targeted electronic transitions have efficiently taken place and significant plasmonic enhancement has been obtained in several configurations of PlasPL and PlasRaman. Future studies will establish practical values of PlasPL and PlasRaman.

As for PlasCat and PlasNLO, the progress is relatively less advanced at the present stage. However, it does not seem that the studies are less promising. The electronic transitions are more complicated than PL processes and need more advanced strategy for the plasmonic enhancement.

Also, we specified the application. Related to the application, IR emitters were addressed; besides, the potential roles of plasmonic resonators at mid-IR ranges were discussed.

Let us conclude this chapter with listing the concrete topics, which would enable the readers to realize the diverse potential of plasmonic resonators.

- PlasPL
- Related to PlasPL, Purcell effect and SPASER were addressed
- Strong-coupling of plasmons with other resonances such as excitons or waveguide resonances.
- PlasRaman
- PlasCat
- PlasNLO
- Plasmon-enhanced light emitters, PV devices, and photodetectors
- As an application of plasmonic PA rather than Plas*, IR emitters at mid-IR ranges were addressed. In addition, the roles of plasmonic resonators at mid-IR ranges were discussed.

References

Adato, R., Yanik, A. A., Amsden, J. J., Kaplan, D. L., Omenetto, F. G., Hong, M. K., Erramilli, S., and Altug, H. (2009). Ultra-sensitive vibrational spectroscopy of protein monolayers with plasmonic nanoantenna arrays, *Proc. Natl. Acad. Sci. U.S.A.* **106**, 19227–19232.

Agrawal, G. P. (2007). *Nonliner Fiber Optics*, 4th edn. (Academic Press, Burlington, MA).

Akselrod, G. M., Argyropoulos, C., Hoang, T. B., Ciracì, C., Fang, C., Huang, J., Smith, D. R., and Mikkelsen, M. H. (2014). Probing the mechanisms of large purcell enhancement in plasmonic nanoantennas, *Nature Photon.* **8**, 835–840.

Albrecht, M. G., and Creighton, J. A. (1977). Anomalously intense Raman spectra of pyridine at a silver electrode, *J. Am. Chem. Soc.* **99**, 5215–5217.

Anderson, M. S. (2000). Locally enhanced Raman spectroscopy with an atomic force microscope, *Appl. Phys. Lett.* **76**, 3130–3132.

Atwater, H. A., and Polman, A. (2010). Plasmonics for improved photovoltaic devices, *Nature Mater.* **9**, 205–213.

Bauch, M., Toma, K., Toma, M., Zhang, Q., and Dostalek, J. (2014). Plasmon-enhanced fluorescence biosensors: a review, *Plasmonics* **9**, 781–799.

Bloembergen, N. (1996). *Nonliner Optics*, 4th edn. (World Scientific, Singapore).

Bloembergen, N., Chang, R. K., Jha, S. S., and Lee, C. H. (1968). Optical second-harmonic generation in reflection from media with inversion symmetry, *Phys. Rev.* **174**, 813–822.

Boyd, R. W. (2008). *Nonliner Optics*, 3rd edn. (Academic Press, Burlington, MA).

Chance, R. R., Miller, A. H., Prock, A., and Silbey, R. (1975). Fluorescence and energy transfer near interfaces: The complete and quantitative description of the Eu^{+3}/mirror systems, *J. Chem. Phys.* **63**, 1589–1595.

Chance, R. R., Prock, A., and Silbey, R. (1978). Molecular fluorescence and energy transfer near interfaces, *Adv. Chem. Phys.* **37**, 1–65.

Choi, B., Iwanaga, M., Miyazaki, H. T., Sugimoto, Y., Ohtake, A., and Sakoda, K. (2015). Overcoming metal-induced fluorescence quenching on plasmo-photonic metasurfaces coated by a self-assembled monolayer, *Chem. Commun.* **51**, 11470–11473.

Choi, B., Iwanaga, M., Ochiai, T., Miyazaki, H. T., Sugimoto, Y., and Sakoda, K. (2014). Subnanomolar fluorescent-molecule sensing by guided resonances on nanoimprinted silicon-on-insulator substrates, *Appl. Phys. Lett.* **105**, 201106.

Christ, A., Tikhodeev, S. G., Gippius, N. A., Kuhl, J., and Giessen, H. (2003). Waveguide-plasmon polaritons: Strong coupling of photonic and electronic resonances in a metallic photonic crystal slab, *Phys. Rev. Lett.* **91**, 183901.

Christopher, P., Xin, H., Marimuthu, A., and Linic, S. (2012). Singular characteristics and unique chemical bond activation mechanisms of photocatalytic reactions on plasmonic nanostructures, *Nature Mater.* **11**, 1044–1050.

Ciracì, C., Poutrina, E., Scalora, M., and Smith, D. R. (2012). Origin of second-harmonic generation enhancement in optical split-ring resonators, *Phys. Rev. B* **85**, 201403(R).

Ding, W., Zhou, L., and Chou, S. Y. (2014). Enhancement and electric charge-assisted tuning of nonlinear light generation in bipolar plasmonics, *Nano Lett.* **14**, 2822–2830.

Dutta, S. K., Mehetor, S. K., and Pradhan, N. (2015). Metal semiconductor heterostructures for photocatalytic conversion of light energy, *J. Phys. Chem. Lett.* **6**, 936–944.

Fischer, U. C., and Pohl, D. W. (1989). Observation of single-particle plasmons by near-field optical microscopy, *Phys. Rev. Lett.* **62**, 458–461.

Fleischmann, M., Hendra, P. J., and McQuillan, A. J. (1974). Raman spectra of pyridine adsorbed at a silver electrode, *Chem. Phys. Lett.* **26**, 163–166.

Fu, C.-C., Ossato, G., Long, M., Digman, M. A., Gopinathan, A., Lee, L. P., Gratton, E., and Khine, M. (2010). Bimetallic nanopetals for thousand-fold fluorescence enhancements, *Appl. Phys. Lett.* **97**, 203101.

Fujishima, A., and Honda, K. (1972). Electrochemical photolysis of water at a smiconductor electrode, *Nature* **238**, 37–38.

Fujishima, A., Zhang, X., and Tryk, D. A. (2008). TiO_2 photocatalysis and related surface phenomena, *Surf. Sci. Rep.* **63**, 515–582.

Fujita, M., Takahashi, S., Tanaka, Y., Asano, T., and Noda, S. (2005). Simultaneous inhibition and redistribution of spontaneous light emission in photonic crystals, *Science* **308**, 1296–1298.

Glass, A. M., Liao, P. F., Bergman, J. G., and Olson, D. H. (1980). Interaction of metal particles with adsorbed dye molecules: absorption and luminescence, *Opt. Lett.* **5**, 368–370.

Goy, P., Raimond, J. M., Gross, M., and Haroche, S. (1983). Observation of cavity-enhanced single-atom spontaneous emission, *Phys. Rev. Lett.* **50**, 1903–1906.

Greffet, J.-J., and Nieto-Vesperinas, M. (1998). Field theory for generalized bidirectional reflectivity: derivation of helmholtz's reciprocity principle and kirchhoff's law, *J. Opt. Soc. Am. A* **15**, 2735–2744.

Hakala, T. K., Toppari, J. J., Kuzyk, A., Pettersson, M., Tikkanen, H., Kunttu, H., and Törmä, P. (2009). Vacuum Rabi splitting and strong-coupling dynamics for surface-plasmon polaritons and rhodamine 6G molecules, *Phys. Rev. Lett.* **103**, 053602.

Hayazawa, N., Inouye, Y., Sekkat, Z., and Kawata, S. (2000). Metallized tip amplification of near-field Raman scattering, *Opt. Commun.* **183**, 333–336.

Hesketh, P. J., Zemel, J. N., and Gebhart, B. (1986). Organ pipe radiant modes of periodic micromachined silicon surfaces, *Nature* **324**, 549–551.

Hobson, P. A., Barnes, W. L., Lidzey, D. G., Gehring, G. A., Whittaker, D. M., Skolnick, M. S., and Walker, S. (2002). Strong exciton–photon coupling in a low-Q all-metal mirror microcavity, *Appl. Phys. Lett.* **81**, 3519–3521.

Hulet, R. G., Hilfer, E. S., and Kleppner, D. (1985). Inhibited spontaneous emission by a rydberg atom, *Phys. Rev. Lett.* **55**, 2137–2140.

Hutchison, J. A., Schwartz, T., Genet, C., Devaux, E., and Ebbesen, T. W. (2012). Modifying chemical landscapes by coupling to vacuum fields, *Angew. Chem. Int. Ed.* **51**, 1592–1596.

Inoue, T., Zoysa, M. D., Asano, T., and Noda, S. (2014). Realization of dynamic thermal emission control, *Nature Mater.* **13**, 928–931.

Iwanaga, M., and Choi, B. (2015). Heteroplasmon hybridization in stacked complementary plasmo-photonic crystals, *Nano Lett.* **15**, 1904–1910.

Iwanaga, M., Choi, B., Miyazaki, H. T., Sugimoto, Y., and Sakoda, K. (2015). Large-area resonance-tuned metasurfaces for on-demand enhanced spectroscopy, *J. Nanomater.* **2015**, 507656.

Jeanmaire, D. L., and van Duyne, R. P. (1977). Heterocyclic, aromatic, and aliphatic amines adsorbed on the anodized silver electrode, *J. Electroanal. Chem.* **84**, 1–20.

John, S. (1987). Strong localization of photons in certain disordered dielectric superlattices, *Phys. Rev. Lett.* **58**, 2486–2489.

Kawata, S., Inouye, Y., and Verma, P. (2009). Plasmonics for near-field nano-imaging and superlensing, *Nature Photon.* **3**, 388–394.

Khajavikhan, M., Simic, A., Katz, M., Lee, J. H., Slutsky, B., Mizahi, A., Lomakin, V., and Fainman, Y. (2012). Thresholdless nanoscale coaxial lasers, *Nature* **482**, 204–207.

Kim, S., Jin, J., Kim, Y.-J., Park, I.-Y., Kim, Y., and Kim, S.-W. (2008). High-harmonic generation by resonant plasmon field enhancement, *Nature* **453**, 757–760.

Kinkhabwala, A., Yu, Z., Fan, S., Avlasevich, Y., Müllen, K., and Moerner, W. E. (2009). Large single-molecule fluorescence enhancements produced by a bowtie nanoantenna, *Nature Photon.* **3**, 654–657.

Klein, M. W., Enkrich, C., Wegener, M., and Linden, S. (2006). Second-harmonic generation from magnetic metamaterials, *Science* **313**, 502–504.

Kneipp, K., Wang, Y., Kneipp, H., Perelman, L. T., Itzkan, I., Dasari, R. R. and Feld, M. S. (1997). Single molecule detection using surface-enhanced Raman scattering (SERS), *Phys. Rev. Lett.* **78**, 1667–1670.

Knight, M. W., Sobhani, H., Nordlander, P., and Halas, N. J. (2011). Photodetection with active optical antennas, *Science* **332**, 702–704.

Lakowicz, J. R., Ray, K., Chowdhury, M., Szmacinski, H., Fu, Y., Zhang, J., and Nowaczyk, K. (2008). Plasmon-controlled fluorescence: a new paradigm in fluorescence spectroscopy, *Analyst* **133**, 1308–1348.

Lehr, D., Reinhold, J., Thiele, I., Hartung, H., Dietrich, K., Menzel, C., Pertsch, T., Kley, E.-B., and Tünnermann, A. (2015). Enhancing second harmonic generation in gold nanoring resonators filled with lithium niobate, *Nano Lett.* **15**, 1025–1030.

Li, J. F., Huang, Y. F., Ding, Y., Yang, Z. L., Li, S. B., Zhou, X. S., Fan, F. R., Zhang, W., Zhou, Z. Y., Wu, D. Y., Ren, B., Wang, Z. L., and Ti, Z. Q. (2010). Shell-isolated nanoparticle-enhanced Raman spectroscopy, *Nature* **464**, 392–395.

Lin, S. Y., Moreno, J., and Fleming, J. G. (2003). Three-dimensional photonic-crystal emitter for thermal photovoltaic power generation, *Appl. Phys. Lett.* **83**, 380–382.

Liu, L., and Chen, X. (2014). Titanium dioxide nanomaterials: Self-structural modifications, *Chem. Rev.* **114**, 9890–9918.

Loudon, R. (2000). *The Quantum Theory of Light*, 3rd edn. (Oxford University Press, Oxford, UK).

Love, J. C., Estroff, L. A., Kriebel, J. K., Nuzzo, R. G., and Whitesides, G. M. (2005). Self-assembled monolayers of thiolates on metals as a form of nanotechnology, *Chem. Rev.* **105**, 1103–1170.

Lozano, G., Louwers, D. J., Rodríguez, S. R. K., Murai, S., Jansen, O. T. A., Verschuuren, M. A., and Rivas, J. G. (2013). Plasmonics for solid-state lighting: enhanced excitation and directional emission of highly efficient light sources, *Light Sci. Appl.* **2**, e66.

Melentiev, P. N., Afanasiev, A. E., Kuzin, A. A., Baturin, A. S., and Balykin, V. I. (2013). Giant optical nonlinearity of a single plasmonic nanostructure, *Opt. Express* **21**, 13896–13905.

Miyazaki, H. T., Kasaya, T., Oosato, H., Sugimoto, Y., Choi, B., Iwanaga, M. and Sakoda, K. (2015). Ultraviolet-nanoimprinted packaged metasurface thermal emitters for infrared CO_2 sensing, *Sci. Technol. Adv. Mater.* **16**, 035005.

Mubeen, S., Lee, J., Singh, N., Krämer, S., Stucky, G. D., and Moskovits, M. (2013). An autonomous photosynthetic device in which all charge carriers derive from surface plasmons, *Nature Nanotechnol.* **8**, 247–251.

Mubeen, S., Zhang, S., Kim, N., Lee, S., Krämer, S., Xu, H., and Moskovits, M. (2012). Plasmonic properties of gold nanoparticles separated from a gold mirror by an ultrathin oxide, *Nano Lett.* **12**, 2088–2094.

Neubrech, F., Pucci, A., Cornelius, T. W., Karim, S., García-Etxarri, A. and Aizpurua, J. (2008). Resonant plasmonic and vibrational coupling in a tailored nanoantenna for infrared detection, *Phys. Rev. Lett.* **101**, 157403.

Nie, S., and Emory, S. R. (1997). Probing single molecules and single nanoparticles by surface-enhanced Raman scattering, *Science* **275**, 1102–1106.

Noginov, M. A., Zhu, G., Belgrave, A. M., Bakker, R., Shalaev, V. M., Narimanov, E. E., Stout, S., Hertz, E., Suteewong, T., and Wiesner, U. (2009). Demonstration of a spaser-based nanolaser, *Nature* **460**, 1110–1112.

Novotny, L., and Hecht, B. (2012). *Principles of Nano-Optics*, 2nd edn. (Cambridge University Press, New York).

Ohtaka, K., and Tanabe, Y. (1986). Photonic band using vector spherical waves. i. various properties of bloch electric fields and heavy photons, *J. Phys. Soc. Japan* **65**, 2265–2275.

Okamoto, K., Niki, I., Shvartser, A., Narukawa, Y., Mukai, T., and Scherer, A. (2004). Surface-plasmon-enhanced light emitters based on ingan quantum wells, *Nature Mater.* **3**, 601–605.

Otto, A., Mrozek, I., Grabhorn, H., and Akemann, W. (1992). Surface-enhanced Raman scattering, *J. Phys.: Condens. Matter* **4**, 1143–1212.

Oulton, R. F., Sorger, V. J., Genov, D. A., Pile, D. F. P., and Zhang, X. (2008). A hybrid plasmonic waveguide for subwavelength confinement and long-range propagation, *Nature Photon.* **2**, 496–500.

Oulton, R. F., Sorger, V. J., Zentgraf, T., Ma, R.-M., Gladden, C., Dai, L., Bartal, G., and Zhang, X. (2009). Plasmon lasers at deep subwavelength scale, *Nature* **461**, 629–632.

Palik, E. D. (1991). *Handbook of Optical Constants of Solids II* (Academic, San Diego, USA).

Pockrand, I., Brillante, A., and Möbius, D. (1982). Exciton–surface plasmon coupling: An experimental investigation, *J. Chem. Phys.* **77**, 6289–6295.

Punj, D., Mivelle, M., Moparthi, S. B., van Zanten, T. S., Rigneault, H., van Hulst, N. F., García-Parajó, M. F., and Wenger, J. (2013). A plasmonic 'antenna-in-box' platform for enhanced single-molecule analysis at micromolar concentrations, *Nature Nanotech.* **8**, 512–516.

Purcell, E. M. (1946). Spontaneous emission probabilities at radio frequencies, *Phys. Rev.* **69**, p. 681.

Puscasu, I., and Schaich, W. L. (2008). Narrow-band, tunable infrared emission from arrays of microstrip patches, *Appl. Phys. Lett.* **92**, 233102.

Puurunen, R. L. (2005). Surface chemistry of atomic layer deposition: A case study for the trimethylaluminum/water process, *J. Appl. Phys.* **97**, 121301.

Rakić, A. D., Djurušić, A. B., Elazar, J. M., and Majewski, M. L. (1998). Optical properties of metallic films for vertical-cavity optoelectronic devices, *Appl. Opt.* **37**, 5271–5283.

Rose, A., Huang, D., and Smith, D. R. (2011). Controlling the second harmonic in a phase-matched negative-index metamaterial, *Phys. Rev. Lett.* **107**, 063902.

Russell, K. J., Liu, T.-L., Cui, S., and Hu, E. L. (2012). Large spontaneous emission enhancement in plasmonic nanocavities, *Nature Photon.* **6**, 459–462.

Schlather, A. E., Large, N., Urban, A. S., Nordlander, P., and Halas, N. J. (2013). Near-field mediated plexcitonic coupling and giant Rabi splitting in individual metallic dimers, *Nano Lett.* **13**, 3281–3286.

Schmelzeisen, M., Zhao, Y., Klapper, M., Müllen, K., and Kreiter, M. (2010). Fluorescence enhancement from individual plasmonic gap resonances, *ACS Nano* **4**, 3309–3317.

Selig, O., Siffels, R., and Rezus, Y. L. A. (2015). Ultrasensitive ultrafast vibrational spectroscopy employing the near field of gold nanoantennas, *Phys. Rev. Lett.* **114**, 233004.

Shen, Y. R. (2003). *The Principles of Nonliner Optics* (Wiely-Interscience, Hoboken, NJ).

Simon, H. J., Mitchell, D. E., and Watson, J. G. (1974). Optical second-harmonic generation with surface plasmons in silver films, *Phys. Rev. Lett.* **33**, 1531–1534.

Soref, R. (2010). Mid-infrared photonics in silicon and germanium, *Nature Photon.* **4**, 495–497.

Stöckle, R. M., Suh, Y. D., Deckert, V., and Zenobi, R. (2000). Nanoscale chemical analysis by tip-enhanced Raman spectroscopy, *Chem. Phys. Lett.* **318**, 131–136.

Takahashi, Y., Inui, Y., Chihara, M., Asano, T., Terawaki, R., and Noda, S. (2013). A micrometre-scale Raman silicon laser with a microwatt threshold, *Nature* **498**, 470–474.

Utikal, T., Zentgraf, T., Paul, T., Rockstuhl, C., Lederer, F., Lippitz, M., and Giessen, H. (2011). Towards the origin of the nonlinear response in hybrid plasmonic systems, *Phys. Rev. Lett.* **106**, 133901.

Väkeväinen, A. I., Moerland, R. J., Rekola, H. T., Eskelinen, A.-P., Martikainen, J.-P., Kim, D.-H., and Törmä, P. (2014). Plasmonic surface lattice resonances at the strong coupling regime, *Nano Lett.* **14**, 1721–1727.

Yablonovitch, E. (1987). Inhibited spontaneous emission in solid-state physics and electronics, *Phys. Rev. Lett.* **58**, 2059–2062.

Ye, Y.-H., Jiang, Y.-W., Tsai, M.-W., Chang, Y.-T., Chen, C.-Y., Tzuang, D.-C., Wu, Y.-T., and Lee, S.-C. (2008). Localized surface plasmon polaritons in Ag/SiO_2/Ag plasmonic thermal emitter, *Appl. Phys. Lett.* **93**, 033113.

Zhang, W., Ding, F., Li, W.-D., Wang, Y., Hu, J., and Chou, S. Y. (2012). Giant and uniform fluorescence enhancement over large areas using plasmonic nanodots in 3D resonant cavity nanoantenna by nanoimprinting, *Nanotechnol.* **23**, 225301.

Zhang, Y., Zhen, Y.-R., Neumann, O., Day, J. K., Nordlander, P., and Halas, N. J. (2014). Coherent anti-stokes Raman scattering with single-molecule sensitivity using a plasmonic fano resonance, *Nat. Commun.* **5**, 4424.

Zhou, L., Ding, F., Chen, H., Ding, W., Zhang, W., and Chou, S. Y. (2012). Enhancement of immunoassay's fluorescence and detection sensitivity using three-dimensional plasmonic nano-antenna-dots array, *Anal. Chem.* **84**, 4489–4495.

Chapter 6

Future Prospects

6.1 Status After two Decades Since the Era of Nanotechnology

6.1.1 *Progress Achieved to Date*

Nanotechnology has been extensively pushed forward in these two decades. As one of the subfields, plasmon-related research area including plasmonics and metamaterials (MMs) have made great, rapid, and diverse progress. This book has described the progress, most of which would not achieved without the investigations on plasmonic resonators (i.e., resonant metallic nanostructures). Among the progress, outstanding development is summarized as follows.

(i) Electromagnetic (EM)-wave propagations and manipulations at subwavelength dimensions. The concrete results have been described in Chapters 1 and 3.

- Extraordinary transmission (Section 1.4.2).
- plasmonic waveguides of metal-insulator-metal (MIM) structures (Section 3.1) and the periodically modulated structures, which were termed fishnet MMs, perfect

Plasmonic Resonators: Fundamentals, Advances, and Applications
Masanobu Iwanaga

ISBN 978-981-4745-06-2 (Hardcover), 978-981-4745-07-9 (eBook)
www.panstanford.com

absorbers (PAs), and IR thermal emitters (Sections 3.6.5, 3.9, and 5.10, respectively).

- subwavelength optical devices, which are able to handle light waves within one vacuum wavelength as a consequence of their high manipulation capability (Sections 3.6.4 and 3.8.2).
- metasurfaces, which are mostly single-layer structures but the periodicity is not limited to the subwavelength dimensions. The complex unit cells increase degree of freedom to manipulate polarization, phase, and wavefront of light waves (Section 1.4.4).

(ii) Plasmonic-enhancement effects have been elaborated in Chapter 5. In particular, the two followings seem to have reached almost maximized enhancement as consequences of the numerous studies over 30 years.

- PlasPL (Section 5.3)
- PlasRaman (Section 5.6)

Individual subjects related to the (i) and (ii) above will turn to the practical applications in the near future.

We here note that many other plasmon-related issues have been described in this book. They do not yet seem to reach satisfactory status though the studies on them already have resulted in a quite large number. This is most likely to mean that there is still room to explore the issues.

6.1.2 *Comparison of Plasmonic Resonators with Other Materials or Artificial Structures*

Figure 6.1 summarizes advantages and disadvantages in plasmonic resonators, comparing them with other materials and artificial structures. Open circles, triangles, and crosses mean that the practical implementations are possible, difficult, and almost no hope (or impractical), respectively. Double open circles mean that the plasmonic resonators or other materials are highly suitable.

Enhancement control As for the enhancement control of electronic transitions in given materials, plasmonic resonators and other

	Plasmonic resonators	Other materials *etc*
Enhancement control	○	○ (PhCs)
Feasibility in design	○	○
Tunability	○	△
Compatibility	○	△
Ultrafast response	◎	×
Low loss	×	◎ (Optical Fibers)
Total evaluation	○	△

Figure 6.1 Classification for advantages and disadvantages in plasmonic resonators. Open circles mean "possible", triangles "difficult" though not impossible, and crosses "almost no hope" or "impractical." Double open circles designate "highly suitable."

dielectric artificial structures like photonic crystals (PhCs) are competing though the former seems to be relatively better in PlasPL and PlasRaman (Sections 5.3 and 5.6).

Feasibility in designs As for feasibility in designs, it is possible to implement the designs in both plasmonic resonators and other artificial structures though the intuitive designs for plasmonic resonators are relatively easy; for example, PAs were frequently designed based on MIM structures (Section 3.9).

Tunability This term means how easy the resonant frequencies (or wavelengths) can be changed and tuned. Many plasmonic resonators have been shown in this book and the resonant frequencies are varied by tuning the metallic nanostructures; for example, Figs. 3.12, 3.52, and 3.61 show the systematic tunability. Even in the metallic nanoparticles, the resonant frequencies were changed in a distinct manner (Fig. 3.5); this evident change stimulated many researches on the nanoparticle plasmonic resonators (Section 3.2).

Compatibility This term means how many materials or structures can efficiently incorporated with. As was described in Chapter 5, the plasmonic resonators have various partners; the diversity is one of the features in plasmonic resonators, making them unique and compatible platforms. In other materials etc., such diverse incorporations have not been reported so far.

Ultrafast response One of the features in plasmonic resonators can be found in the ultrafast response. It directly comes from line width $\Delta\omega$ of the resonances, which are characterized using absorption spectra or extinction spectra.[a] Lifetime of the plasmonic resonance τ_{pl} is estimated to be $\tau_{\mathrm{pl}} \sim 1/\Delta\omega$, which is typically 10~30 fs.

Experimentally, it was found that radiative SPP on a shallow grating structure has lifetime of 30 fs while an asymmetric SPP mode that suppresses radiation loss has relatively longer lifetime of 200 fs (Ropers et al., 2005). This example suggests that ultrafast responses in plasmonic resonators can be varied according to the choices.

Such ultrafast responses from the ground state via excited states to the ground state are less frequently found in electric-dipole transitions in semiconductors and insulators. The ultrashort lifetime implies that, if one is able to incorporate resonance in a plasmonic resonator with a targeted electronic transition in the partner material, an ultrafast response can be obtained from the plasmon-enhanced transition. PlasNLO (Section 5.8) will substantially develop along this direction.

Low loss As for low loss, plasmonic resonances are often associated with large extinction; therefore, optical loss is inevitable and prevents long-distance propagation at the resonant frequencies. Thus, the size of plasmonic resonators in practical usage is limited to very small, typically, subwavelength dimensions.

[a]Extinction is proportional to absorption efficient in a simple configuration and therefore is often used as substitution of the absorption efficient.

The small interaction domains have been frequently addressed. For example, the highly efficient polarizer of stacked complementary (SC) structure (Section 3.8.2) and perfect emittance plasmo-photnic (PlasPh) crystals demonstrating more than 2600-fold FL-intensity enhancement (Section 5.3) were realized in the plasmonic resonators of subwavelength thickness; in both cases, the optical losses were roughly 15% and were not obstacles for the characteristic features.

On the other hand, optical low loss has been established in optical fibers in an extremely efficient manner; no better materials regarding optical loss will not appear on the earth.

Total evaluation Overall, total evaluation has been judged, taking all the criterions above into account. Plasmonic resonators have various advantages and a disadvantage, i.e., optical loss; however, as far as the structures are kept in subwavelength dimensions, the disadvantage will be compensated by the advantages.

In contrast, other materials and artificial structures can serve as good materials or devices for a specific purpose, whereas they are less suitable to explore diverse usage and application. For example, optical fibers have a strong advantage in the low-loss propagation of light while they have a disadvantage in compatibility because of the restriction in shape.

Unsuitable issues tackled with plasmonic resonators Considering the progress in the plasmonic resonators over tens of years, it is inevitable to mention that some issues turned out to be unsuitable for the plasmonic resonators. In particular, we mention the followings.

- Usage of plasmonic waveguides for long-distance propagation. As is well known, propagation length of SPP is at most hundreds microns even on flat crystalline Ag surface. It is, unfortunately, reasonable to call them lossy waveguides.
- Application that requires low-loss transport of EM-wave power (or intensity). This is related to the previous comment on the plasmonic waveguides. Ideally low-loss and long-distance transport obtained in quartz optical fibers is not expected for the plasmonic resonators.

- Usage of plasmonic resonators instead of transparent materials. It has been sometimes intended to obtain low-loss plasmonic resonators from the similar motives to the two comments just above. However, plasmonic resonators mostly have large extinction, linked to large light absorption. The absorbed light is converted very efficiently into the resonant modes. Thus, it is impractical to replace transparent materials with plasmonic resonators. Instead, other usages should be considered.

Readers just realize that the unsuitable issues are connected to one motive; that is, to replace lightguide of almost ideally transparent materials with any plasmonic resonator. It is, of course, possible to reduce the large loss by plasmonic resonators with incorporating some gain materials (Wuestner et al., 2010; Xiao et al., 2010); still, considering that the current long-distance lightguides in practical use by optical fibers were already associated with the amplification subsystem, a conclusion will come that it is extremely hard to incorporate any plasmonic resonator for the long-distance EM-wave propagation. Therefore, the trials to intend the replacement will end in unsuccessful results.

Through the discussion on the unsuitable issues, we intend to stress that plasmonic resonators should be exploited in sub-wavelength dimensions for realizing significant enhancing effects or efficient EM-wave manipulations. In such situation, it becomes evident that the enhancement effect and the manipulation capability are more prominent than the optical loss by the plasmonic resonators.

6.2 Directions Being Opened

In this section, we survey ongoing and further developing subjects on plasmonic resonators.

Advanced elaborated designs One of the recent progress is found in metasurfaces of single-layer structures. The new twist was to introduce complex unit cell (Yu et al., 2011); a schematic was shown in Fig. 1.22. In MMs that are subwavelength periodicity, the unit cells were limited to quite simple ones especially at optical frequencies, as shown in Figs. 1.20 and 3.19. In contrast, metasurfaces consist of complex unit cell based on an empirical design and have degree of freedom to control wavefront using diffractive harmonics; as a result, achromatic wavefront control became possible (Aieta et al., 2012, 2015).

As for advanced designs for photonic structures including plasmonic resonators, unempirical algorithms such as genetic algorithm (GA) have been reported so far (e.g., Shen et al., 2003; Goh et al., 2007; Iwanaga, 2009). Still, the reports, in which 3D structures were adequately searched, are hardly found because huge computer resources are needed. However, metasurfaces are essentially 2D structures and can be adequately explored by the unempirical algorithms like GA. Thus, novel and useful designs will be expected to be found in the near future.

Improvement of quality of fabricated metallic nanostructures The trials to pursue the better quality of metallic thin films were continued, succeeding in obtaining single-crystalline Ag thin films by molecular beam epitaxy (MBE) growth (Park et al., 2012; Lu et al., 2012; Wu et al., 2014). This means that the best-quality Ag, which is associated with the minimum optical loss, can be obtained in the nanostructures. Still, note that, even when the MBE technique is introduced, the lowest loss Ag films are limited to the single-crystal quality in literature (Palik, 1991).[b]

Making use of the high-quality Ag film, surface-plasmon-amplified stimulated emission resonator (SPASER) was operated at a very low threshold (Lu et al., 2012). Also, subwavelength metallic groove of simple Ag–air periodic structure, which is a MM, was fabricated and made possible to move the hyperbolic-dispersion wavelengths to the visible range (High et al., 2015). These results mean that the present lithographic techniques are able to produce Ag nanostructures with the high quality.

Trials to find new plasmonic materials It has been a frequent asked question whether any other materials serve as plasmonic materials instead of the noble metals. This question was already replied in an extensive way to examine permittivity (or refractive index) in

[b]We note that Ag appears in the volume III.

diverse materials such as graphene, superconductors, alkali metals, TiN, ZrN, Ta, W, and so on (Boltasseva and Atwater, 2011; Tassin et al., 2012; Guler et al., 2014).

To overcome the disadvantage in the optical loss of the noble metals, doped semiconductors and refractory materials like TiN were explored as alternatives. This attempt is interpreted as a material parameter search, just as MMs were initially expected to realize a new domain of material parameters that is out of reach for known condensed matters.

It has eventually turned out that the noble metals are one of the best solutions in the visible and NIR ranges. On the other hand, it was found that, at the IR ranges, the refractory materials are able to serve as plasmonic materials instead of the noble metals in terms of the material parameter, i.e., permittivity.

One of the advantages in the refractory materials is to be able to stable at high temperatures (Li et al., 2014). Therefore, they have a potential to serve as good materials for high-temperature light absorbers and IR emitters. Other possibilities such as various plasmonic-enhancement effects discussed in Chapter 5 have not been explored for the alternative materials so far. For example, Au has been exploited for PlasRaman, because of the chemical stability and the surface state. It is not obvious that such material properties are realized in the alternatives, even though permittivity is similar. In more practical stages for actual application, the feasibility in fabrication and the compatibility discussed in Fig. 6.1 will be issues to be examined.

6.3 Challenges in Near Future

There are several issues to deserve exploring though they seem to be hardly pursued now. We here address the issues from a general point of view.

(i) Making optical devices ultracompact.
One of the features in plasmonic resonators is that they work even in subwavelength dimensions. Therefore, optical devices

of plasmonic resonators, in principle, can be ultracompact. This feature is hard to be realized by other materials etc.

Moreover, the ultracompact devices are able to operate with very small power. As a result, assembled circuits of the ultracompact devices will operate with smaller power than the existing devices of the equivalent functions.

(ii) Plasmonic resonators of ideally high efficiency.
For any purpose, the goal should be set to ideally high efficiency, which makes the values of the research and development concerning plasmonic resonators evident and unique.

(iii) Plasmonic resonators realizing unique functions.
The feasibility in design and compatibility make it easy for plasmonic resonators to have unique functions.

The individual discussions on Plas* in Chapter 5 suggested that plasmonic resonators are competing regarding some issues with other materials or artificial structures such as PhCs. Therefore, it is not a good strategy to choose competing subjects as a theme making use of plasmonic resonators. Of course, if the plasmonic resonators are expected to reach ideally high efficiency, the study should be kept continued until the aim gets accomplished.

Generally, it is a good strategy in the research on plasmonic resonators to select themes that provide unique roles to plasmonic resonators. When ultracompactness, ideally high efficiency, and/or ultrafast response are expected, plasmonic resonators will be unique solutions.

(iv) Quest for functional units including plasmonic resonators.
Plasmonic resonators studied to date are, frankly speaking, elements. By assembling them in systematic designs, functional units or instruments will be produced. As various functions are being realized, the real values of plasmonic resonators will manifest themselves.

6.4 Concluding Remarks

This chapter addressed future prospects in plasmonic resonators. Thus, we do not intend to state conclusions of this book but have tried to provide a view (or views) for possible directions in the near future.

Summing up the piled-up results and the current understandings regarding the plasmonic resonators, we believe that here is not the goal but is a milestone for future development. Thus it is worth noting the status from a frank point of view. It will be our great pleasure if we can stimulate future achievement with the contents and implications in this book. Significant development on plasmonic resonators would contribute to epoch-making innovations in the near future. Personally, it will be one of the greatest pleasures to add the advances to this book in the future.

References

Aieta, F., Genevet, P., Kats, M. A., Yu, N., Blanchard, R., Gaburro, Z., and Capasso, F. (2012). Aberration-free ultrathin flat lenses and axicons at telecom wavelengths based on plasmonic metasurfaces, *Nano Lett.* **12**, 4932–4936.

Aieta, F., Kats, M. A., Genevet, P., and Capasso, F. (2015). Multiwavelength achromatic metasurfaces by dispersive phase compensation, *Science* **347**, 1342–1345.

Boltasseva, A., and Atwater, H. A. (2011). Low-loss plasmonic metamaterials, *Science* **331**, 290–291.

Goh, J., Fushman, I., Englund, D., and Vučković, J. (2007). Genetic optimization of photonic bandgap structures, *Opt. Express* **15**, 8218–8230.

Guler, U., Boltasseva, A., and Shalaev, V. M. (2014). Refractory plasmonics, *Science* **344**, 263–264.

High, A. A., Devlin, R. C., Dibos, A., Polking, M., Wild, D. S., Perczel, J., de Leon, N. P., Lukin, M. D., and Park, H. (2015). Visible-frequency hyperbolic metasurface, *Nature* **344**, 192–196.

Iwanaga, M. (2009). Optically deep asymmetric one-dimensional metallic grooves: Genetic algorithm approach, *J. Opt. Soc. Am. B* **26**, 1111–1118.

Li, W., Guler, U., Kinsey, N., Naik, G. V., Boltasseva, A., Guan, J., Shalaev, V. M., and Kildishev, A. V. (2014). Refractory plasmonics with titanium nitride: Broadband metamaterial absorber, *Adv. Mater.* **26**, 7959–7965.

Lu, Y.-J., Kim, J., Chen, H.-Y., Wu, C., Dabidian, N., Sanders, C. E., Wang, C.-Y., Lu, M.-Y., Li, B.-H., Qiu, X., Chang, W.-H., Chen, L.-J., Shvets, G., Shih, C.-K., and Gwo, S. (2012). Plasmonic nanolaser using epitaxially grown silver film, *Science* **337**, 450–453.

Palik, E. D. (1991). *Handbook of Optical Constants of Solids II* (Academic, San Diego, USA).

Park, J. H., Ambwani, P., Manno, M., Lindquist, N. C., Nagpal, P., Oh, S.-H., Leighton, C., and Norris, D. J. (2012). Single-crystalline silver films for plasmonics, *Adv. Mater.* **24**, 3988–3992.

Ropers, C., Park, D. J., Stibenz, G., Steinmeyer, G., Kim, J., Kim, D. S., and Lienau, C. (2005). Femtosecond light transmission and subradiant damping in plasmonic crystals, *Phys. Rev. Lett.* **94**, 113901.

Shen, L., Ye, Z., and He, S. (2003). Design of two-dimensional photonic crystals with large absolute band gaps using a genetic algorithm, *Phys. Rev. B* **68**, 035109.

Tassin, P., Koschny, T., Kafesaki, M., and Soukoulis, C. M. (2012). A comparison of graphene, superconductors and metals as conductors for metamaterials and plasmonics, *Nature Photon.* **5**, 259–264.

Wu, Y., Zhang, C., Estakhri, N. M., Zhao, Y., Kim, J., Zhang, M., Liu, X.-X., Pribil, G. K., Alù, A., Shih, C.-K., and Li, X. (2014). Intrinsic optical properties and enhanced plasmonic response of epitaxial silver, *Adv. Mater.* **26**, 6106–6110.

Wuestner, S., Pusch, A., Tsakmakidis, K. L., Hamm, J. M., and Hess, O. (2010). Overcoming losses with gain in a negative refractive index metamaterial, *Phys. Rev. Lett.* **105**, 127401.

Xiao, S., Drachev, V. P., Kildishev, A. V., Ni, X., Chettiar, U. K., Yuan, H.-K., and Shalaev, V. M. (2010). Loss-free and active optical negative-index metamaterials, *Nature* **466**, 735–738.

Yu, N., Genevet, P., Kats, M. A., Aieta, F., Tetienne, J.-P., Capasso, F., and Gaburro, Z. (2011). Light propagation with phase discontinuities: Generalized laws of reflection and refraction, *Science* **334**, 333–337.

Appendix A

Abbreviations and Symbols

A.1 Abbreviations

AFM Atomic force microscope

CD device Circular dichroic device (Sections 3.6.4 and 3.7)

EBL Electron-beam lithography (Section 1.6)

EELS Electron energy-loss spectrum

E field Electric field

ER Electroreflectance (Section 4.1)

EM field Electromagnetic field

FEM Finite element method (Section 1.5.3)

FDTD method Finite-difference time-domain method (Section 1.5.2)

FIB Focused ion beam (Section 1.6)

FL Fluorescence. Luminescence that comes from molecular transition from LUMO to HOMO is called FL. See also PL below.

H field Magnetic field. The H-field vector is usually represented with **H** in Maxwell equations.

IMI Insulator-metal-insulator. IMI structure is a metallic slab in insulator circumstance, described in Section 3.1.2. See also MIM below. Further stacking results in IMIMI structure (Section 3.1.3).

IR Infrared. IR range is a very wide wavelength range from 700 nm to about 100 μm in air, containing NIR, mid-IR, and far-IR ranges. See also NIR and mid-IR below.

mid-IR Mid infrared. A wavelength range from 2 μm to 5 μm is called mid-IR range. Plasmonic resonators at the mid-IR range were produced as perfect absorbers (Section 3.9) and thermal emitters (Section 5.10).

MIM Metal-insulator-metal. MIM structure was one of the popular and oldest structures of plasmonic resonator, described in Section 3.1.1. See also IMI above.

MMs Metamaterials (Section 1.4.3)

NIR Near infrared. Typically, a wavelength range is from 700 to 2000 nm in air though the long-wavelength edge is not clearly defined. Sometimes, NIR is called short-wavelength infrared (SWIR).

PA Perfect absorber. Materials or artificial structures that satisfy 100% light absorption are called PAs. In application for IR emitters and FL-enhancing platforms, PAs are used as perfect emitters. Note that PAs are equivalent to perfect emitters as far as the Kirchhoff's radiation law holds (Section 3.9).

PL Photoluminescence. Although FL is a kind of PL, PL is more general because PL also includes luminescence from semiconductors, ionic crystals, and other materials in addition to organic luminescent molecules. Generally, when luminescence is induced by photoexcitation, the luminescence is called PL.

PlasPh Plasmo-photonic. The notion is introduced in Section 3.8 to describe hybrid resonances of plasmons and photonic guided resonances in the SC structures. See also SC structure below.

Plas* Plasmon-enhanced *. The symbol * can be filled with various terms such as PL, Raman scattering, catalysis, nonlinear optics, and so on; accordingly, PlasPL, PlasRaman, PlasCat, and PlasNLO are described in Chapter 5.

PlasCat Plasmon-enhanced catalysis. See also Plas* just above (Section 5.7)

PlasPL Plasmon-enhanced PL. See also Plas* above (Section 5.3)

PlasRaman Plasmon-enhanced Raman scattering. PlasRaman contains SERS, TERS, and SEIRA. See also Plas*, SERS, TERS, and SEIRA (Section 5.6).

PhC Photonic crystal

PlC Plasmonic crystal

RCWA Rigorous coupled-wave analysis. RCWA is a method to numerically solve Fourier-transformed Maxwell equations, enabling to obtain eigenmodes in a layer of periodicity. RCWA is sometimes called Fourier modal method. One of the key features is to be able to handle metallic nanostructures in a precise way. This book adopts RCWA as a main numerical method together with S matrix (below). The formulation is described in Section 1.5.1.

SAM Self-assembled monolayer. SAM is incorporated to form a nm-precision gap between metallic nanostructures and other materials. The experimental results are described in Section 5.3.1.

SC structure Stacked complementary structure. As for metallic structures, complementary structures are stacked (for example, Figs. 1.26, 3.41, and 3.44). The fabrication procedure is mentioned in Section 1.6. The physical properties of the SC plasmonic resonators are described in Section 3.8.

SEIRA Surface-enhanced infrared absorption. The absorption comes from molecular vibration modes; therefore, SEIRA is discussed in connection Raman scattering though SEIRA itself is, strictly, not Raman scattering (Section 5.6.3).

SEM Scanning electron microscope

SERS Surface-enhanced Raman scattering (Section 5.6.1)

S matrix Scattering matrix. S matrix is a general formalism describing phenomena containing multi-inputs and multi-outputs in physics. In this book, light scattering is mainly described using S matrix because it is free from numerical divergence. See also T matrix below (Section 1.5.1).

SMIM Stratified metal–insulator metamaterials (Section 3.6.2)

SPASER Surface-plasmon-amplified stimulated emission resonators (Section 5.4)

SPP Surface plasmon polariton (Section 1.3)

SR imaging Super-resolution imaging (Section 3.6.2)

SRR Sprit-ring resonator (Section 1.4.3)

TEM Tunneling electron microscope

TERS Tip-enhanced Raman scattering (Section 5.6.2)

T matrix Transfer matrix. T matrix enables us to understand the mathematical formulation for input (or incidence) and outputs (or transmission and reflection). However, T matrix generally includes numerical divergent term(s) and therefore is not useful in numerical implementations except for multilayer structures, so that S matrix mentioned above is normally employed (Sections 1.5.1 and 3.6.2).

UV Ultraviolet

1D One-dimensional

2D Two-dimensional

3D Three-dimensional

A.2 Symbols

Physical quantities are frequently used in this book. We list them as follows.

- **D**: electric flux density (or electric displacement)

- **E**: electric-field vector
- **B**: magnetic flux density (or magnetic displacement)
- **H**: magnetic-field vector
- **A**: vector potential
- **k**: wavevector
- c_0: velocity of light in vacuum, equal to $1/\sqrt{\varepsilon_0 \mu_0}$
- χ: response function or susceptibility
- ε_0: permittivity in vacuum
- ε: relative permittivity
- k_0: wavenumber in vacuum, equal to ω/c_0
- μ_0: permeability in vacuum
- μ: relative permeability, which is set to unity at optical frequencies
- $\hat{n}$: complex refractive index
- ω: angular frequency. $\hbar\omega$ is photon energy

Index